Quantum Ethics

A Spinozist Interpretation
of Lattice Quantum Field Theory

Sébastien Fauvel

Acknowledgements

First and foremost, I would like to thank my wife Angela for having let me enjoy her wisdom and love radiating being for all these years. She made me discover Spinoza's philosophy and this publication would never have been possible without the help of her brilliant intelligence, of her constructive criticism, of all her love and support. Thank you!

I would also like to thank my colleague Markus Matteucci for initiating me into the art of book publishing, my fellow students Christian Hagendorf and Tristan Machado for some fruitful discussions that did much for my understanding of Physics, my teachers Christian Swit and Stéphane Gentil who introduced me to the joys of Philosophy and Mathematics, my uncle Bertrand Liaudet and my regretted "godfather" Frédéric Poupaud for having made me discover the world of Algorithmic and Informatics at a far too early age. They all did very much for my personal evolution.

Preface

This book has been written with the main concern of providing the scientific community with a mean of building a bridge between physicists and philosophers in the field of Quantum Physics. It defines a common language to describe the realm of our experience of the world and I truly hope that this new language will find a large audience in both communities. For physicists, it stresses the importance of developing a well-defined mathematical formalism for Quantum Field Theory, since this is the necessary condition for philosophers to identify the underlying ontology, which builds the base of every philosophical discourse on the implications of Quantum Physics. The development of a coherent, convincing discourse by philosophers builds in turn the ground on which every conceptually and technically correct vulgarization effort can foot, and contributes thereby to broaden the popularity and acceptance of Quantum Physics through the whole society and especially among prospective students. Philosophers, on the other hand, gain a mean of confronting their ideas with the latest insights into fundamental physics, of expressing these ideas in terms of naturalistically grounded metaphysics and of articulating speculative thoughts in the uncertainty zones remaining within the physical theory itself or inherent to it.

The careful reader already familiar with Quantum Physics will find here the very first mathematically well-defined formulation of Quantum Field Theory coming along with an intellectually satisfactory interpretation, perfectly capable of explaining all known quantum phenomena, and I am very pleased to present it to your curiosity today! However, although you will find, in the first edition of this book, all the main ideas I wanted to develop here, there still are a few areas which I couldn't yet find the time to develop to the extend they would have deserved. For instance, I do consider the formulation of Quantum Field Theory presented in this book, which is essentially a lattice regularization, as an acceptable fundamental theory, although it obviously lacks to respect the heuristical principles of Gauge and Poincaré invariance on which any textbook introduction to Quantum Field Theory relies. I included a few calculation examples in order to demonstrate that, at usual energy and distance scales, the resulting physics is the same as expected, but it would have been useful, to convince the skeptical reader, to add a classical renormalization example, a treatment of the weak and strong interactions as well as of the Higgs mechanism and of some gravitational model. The philosophical aspects of this book might have deserved a more extensive treatment, too. I have given a few basic examples to show how the physical theory can be used as a reference language in order to express philosophical questions, but I would have to add more various examples to give an adequate idea of the potential of this method.

I am preparing a second, augmented edition of this work so that the ideas presented here become more explicit and accessible to a broader public. For the time being, I wish you a fruitful and enjoyable reading as well as long hours of delightful meditation!

January, 2013
Sébastien Fauvel

Contents

History of this book

> For one of the most stringent tests of any physical theory
> is the prediction of its own creation process.

Sébastien Fauvel,
Quantum Ethics [8]

How would Quantum Field Theory look like if we stopped for a while developing it further as if it were the draft of a yet-to-be-discovered Theory of Everything, and just started to reformulate the Standard Model as a mathematically and conceptually coherent physical theory? And what would such a theory tell us about the world and about ourselves, which remains hidden in the ill-defined formulations we've grown up with through the last decades? As I started back in 2010 to reflect on these questions, I didn't have yet a clear vision of what this work would lead me to. I just had the feeling that these very basic questions hadn't been interesting anyone any more for a far too long time, and that we should actually have the means by now, with our understanding of Renormalization, of writing down a well-defined Quantum Field Theory reasonably accounting for all known experimental data (excepted General Relativity phenomena) – which means essentially that it has to be compatible with the Standard Model at known energy scales. I was quite confident that I could find a physically sound regularization of the Standard Model, which I simply wouldn't consider as an approximation, but take as the exact theory itself, the Standard Model being an ill-defined idealization of it. The models used in computer simulations of lattice Quantum Chromodynamics, for instance, would show me the way. Of course, I knew that I wouldn't be able to derive the theory from the usual first principles any more, but given that all the attempts of axiomatic Quantum Field Theory to construct well-defined interacting fields upon these first principles had failed miserably, I thought that maybe they could be misleading in the end. Anyhow, I had never been very fond of the heuristical construction of Quantum Field Theory based on Gauge and Poincaré invariance. Developing the whole mathematical apparatus of Representation Theory to simply derive the expression of spin 1 and spin 1/2 spinors as irreducible unitary representations of the Poincaré group had always seemed far too expensive to me, and Gauge transformations mixing particle fields far too artificial to make up a fundamental symmetry of Nature.

So I felt free to redefine the Hilbert space of the quantum states without paying much attention to these first principles and focused instead on the mathematical well-definedness of the theory, and in particular of the Schrödinger equation. The most evident way of insuring a well-defined solution at all times is to make the Hilbert space finite dimensional, which has two major physical implications. The most important one is that the physical space itself, too, has to be finite, *i.e.* to consist in a finite number of points. The simplest way to take this constraint into account is to define space as a finite lattice, like in computer simulations of Quantum Chromodynamics, and to adapt the expression of the Hamiltonian operator of the Standard Model, developed on the momentum basis, by simply using a discrete Fourier transform on the lattice. This formulation of the theory, considered as a fundamental theory and not as a numerical approximation, has evident ontological and cosmological implications. It is interesting to see, for instance, how modern physics addresses thus the atomist polemic of ancient Greece, *i.e.* the question if matter can be, in principle, indefinitely separated into smaller pieces, or if there are smallest building blocks of matter. The answer of this theory were not only that elementary particles are the smallest, point-like building blocks of matter, but that space itself is constituted of smallest, point-like building blocks, and even of a finite number of them! Incidentally, the void between point-like particles imagined by Greek atomists like Democritus acquires a very different quality, too. There is still a notion of void as the unrealized potentiality of the presence of matter, represented by an unoccupied lattice site, but this site, although it is empty of matter, is still something familiar, identifiable, something we could put a name on. Psychologically, the void loses thus much of the threatening quality of the indiscernible. The empty space that we might tend to imagine between the lattice sites isn't actually part of the material world, it is purely virtual and has no physical relevance.

From a cosmological perspective, the finiteness of space is also a very interesting aspect. It addresses the old question of knowing whether there is something like a frontier of the universe or if the universe is infinite, and it offers a very original answer. According to this theory, the universe is both finite and boundless; it actually has a toroidal structure, which is not of topological nature, but reveals itself at the level of the field dynamics: Wave packets will transit smoothly from one side of the finite lattice to the opposite one without experiencing any discontinuity. So the light we emit, for instance, could come back to us from the opposite direction after having traveled through the whole universe. Yes, if the universe were smaller, maybe you could see the Earth looking at the stars... and the position of the closest images of the Earth in the night sky would give you the direction of the lattice axes, by the way.

The second physical implication of the finite dimension of the Hilbert space is the existence of a maximum occupation number for boson fields. I wondered if there were any good theoretical reason to assume an unbounded number of bosons per field mode, and I actually didn't find any. Of course, the commutation relations usually considered as essential properties of the creation operators would break down when the maximum number of particles is being reached, but these relations, relicts of a heuristical construction of Quantum Field Theory based on the harmonic oscillator model of Quantum Mechanics, are not really necessary to define creation operators. In fact, it is quite straight-forward to define a basis of the Hilbert space on a finite lattice, you just have to take as basis vectors field configurations defined as functions giving the number of particles of each kind at each lattice site. And it isn't more complicated either to define creation operators as adding one particle of a given kind at a given lattice site, as long as a given maximum occupation number hasn't been reached. The normalization factors implied by the commutation relations can then be moved to the spinors, where they actually belong. The situation is quite similar for fermions: If you don't construct the Hilbert space heuristically as a Fock space over the one-particle Hilbert space of Quantum Mechanics, the sign factors implied by the anticommutation of the creation operators can be moved to the spinors too. So in the end, there isn't any qualitative distinction to be made between bosons and fermions; the same creation operators can be used in both cases, differing only in their maximum occupation numbers. In fact, if you don't construct the Hilbert space as a Fock space, but define it directly (or use a Fock space modulo particle labels permutations), there is no Spin-Statistics Theorem classifying particles into bosons and fermions according to their spin any more. This famous theorem relates the integer or half-integer character of the spin to the possible sign change happening to the quantum state when the labels of two particles of the same type are being exchanged. But the notion of exchanging the labels of two particles doesn't actually have any physical meaning, it only makes sense in the Fock space formalism, and is a mere mathematical artifact. I think it is important to realize that the Spin-Statistics Theorem, traditionally considered as one of the greatest insights provided by Special Relativity into Quantum Field Theory, actually doesn't have any profound physical meaning, and doesn't establish, as it is often being stated, a connexion between the geometry of space-time and the collective behavior of particles. It only expresses a property of the "unphysical" Fock space formalism, and becomes meaningless as soon as you consider the "physical" quantum states modulo particle labels permutations. So the categories of 'bosons' and 'fermions' are not implied by Special Relativity, as far as their collective behavior is concerned; only the form of the spinors is. Determining experimentally the maximum occupation number for each boson field is still an open question: For "heavy" bosons like the Z boson, for instance, I don't think that a lower bound much greater that one can already be established with current experimental data...

Once I had constructed this well-defined framework for Quantum Field Theory and made a first proof-of-concept by integrating Quantum Electrodynamics, I left the paper draft I had written by that time rest for a while, took care of my newborn son and started reading a book from the Philosophy library of my wife that had been intriguing me for a while: A French translation of Spinoza's *Ethics*. The reading would accompany me through the whole summer of 2011 and make a lasting impression on me. The subtle way Spinoza integrates subjective experience into the physical world reminded me of von Neumann's hypothesis that mind could somehow cause the collapse of the quantum state of a system upon measurement, and I realized that, within the well-defined framework I had constructed, we had the possibility for the first time to give a formally very precise definition of what von Neumann had meant. This would provide a precise answer to the measurement problem, and probably the first one that isn't only psychologically motivated, but also constrained by formal consistency. So I started to figure out how to relate subjective experience to the state of the material world in quantum physical terms, and re-read Spinoza with this question in mind. Following von Neumann's interpretation, I should relate a mental state to a Hilbert subspace in such a way that the Hilbert space be a direct sum of the subspaces corresponding to each possible mental state. Making the assumption that we have to do with different states of a single subjective experience in this decomposition leads directly to the paradox of Wigner's friend, that is, when several bodies (brains?) are present at once – and it is the case most of the time, isn't it? –, which one oughts to determine the mental state and trigger the collapse? Escaping this issue requires to describe the mental state in its totality, *i.e.* to specify the number of subjects having each possible subjective experience at a time, so that a mental state is, basically, described with the same formalism as a field configuration over subjective experiences. And exactly as this is the case for particles in particle fields, subjects are *indistinguishable* at a fundamental level. There is nothing like "my" mind or "your" mind, each one having its own personal history that could, in principle, be tracked back from birth to death. Pretty much like single particles don't have any individual trajectory in Quantum Physics, single subjects don't have any individual history either. As Spinoza would say, we are all thinking together in God; we participate of a single mental reality and don't have any individual existence below this ontological level. This will probably sound crazy to most readers, and it is probably one of the reasons why Spinoza has been excommunicated for heresy in his time. But it is actually an utmost self-consistent point of view, and the only one consistent with Quantum Field Theory so far. I cannot but warmly advise you to take a closer look at *The Ethics*; re-reading Spinoza and seeing how a 17th century heresy meets Quantum Physics is really a very exciting experience. The pantheist thesis of Spinoza fits incredibly well in the world view sustained by Quantum Field Theory; neither your body, enmeshed by quantum entanglement with other ones, nor your mind, indistinguishable from other ones, have any individual existence: Nothing exists but God, aka Nature. This is basically the idea of this book, and given that no other interpretation of Quantum Physics integrates so deeply into the formalism of Quantum Field Theory, this made me think that this book was worth writing it, and I guess it will be a joy for many science philosophers to see that the latest achievements in fundamental physics are leading us back, eventually, from a materialistic to a pantheist philosophy.

As soon as I had developed this Spinozist model of the mental world (which builds up, together with the material world of quantum fields, the physical world as a whole), I got confronted with the old question of the status of time in Quantum Physics. The controversies on this subject have been summarized very concisely by Wolfgang Pauli in his statement that there cannot be any time observable in Quantum Physics. In the Copenhagen interpretation, indeed, time isn't a property of the quantum system under observation; it isn't being measured quantum physically, but *classically*, and correlated with quantum measurement results. When you measure the fluorescence lifetime of ruby, for instance, you only measure the presence of emitted photons on a quantum physical way, which implies the collapse of the system's quantum state, but you measure the time at which the photodetector gets activated by simultaneously reading a clock in a classical way. That is a very strange feature of the quantum/classical dichotomy of the Copenhagen interpretation, and it leaves one very basic question completely open: There is no way to predict quantum physically *when* the quantum measurement process and the collapse of the quantum state will take place, or even to find out the time distribution of the measurement process in a statistical way. The Copenhagen interpretation only defines the statistical distribution of the possible measurement results assuming a measurement is being performed at a given time, but doesn't tell anything about the conditions under which a quantum measurement will actually happen – basically because measuring is considered as an act taking place in the classical world, which escapes quantum physical description. The reason why this uncertainty about the time at which a quantum measurement happens doesn't have any consequences on our ability to derive statistical results from the theory was already clear in the 1930's: As von Neumann pointed out, it wouldn't make any statistical difference if the collapse of the quantum state happened upon an interaction of the quantum system with a measurement apparatus, or upon an interaction of the quantum system including measurement apparatus with the observer, or at any stage inbetween. And even if the observer wouldn't read the output of the measurement apparatus, the interaction of the quantum system with it, like any process introducing a strong correlation of its state with the environment, would yield quantum decoherence effects which are practically impossible to tell apart from the effects of an hypothetical collapse, as far as the statistical measurement results are concerned. So we have practically no means of finding out at which stage the collapse is taking place, and addressing this question remains a purely theoretical issue of no practical interest. Nevertheless, it has to be addressed by any theory going beyond the Copenhagen interpretation and trying to describe collapse as a physical process independent of the free will of the observer, which is subsumed in the classical world view. There are lots of so-called spontaneous collapse theories, developed originally by John Bell followed by many others, which generally describe collapse as a dynamical process, yielding *in fine* to the same states as an abrupt orthogonal projection would do. But these models are purely materialistic and don't address the question of describing subjective experience in physical terms. The suggestion of von Neumann that mind could cause the collapse of the quantum state, which would get projected to quantum states of the brain corresponding to a definite subjective experience, seemed much more promising to me, as I was looking forward to sketching a more comprehensive world view in physics. So I stuck to the rather conservative hypothesis of an abrupt collapse of the quantum state via a random orthogonal projection to one of the Hilbert subspaces corresponding to a given mental state, and I had to define precisely when this process would happen. In doing so, you are totally free as a theoretical physicist, because, as I said before, collapse and quantum decoherence have practically the same signature in statistical measurement results, so that we can never be sure of having observed a collapse or not. I rejected the hypothesis of a continuous collapse, because continuous stochastic processes are only idealizations, so I supposed rather that collapses happen at discrete times. This implies that our mental state evolves discontinuously, although we usually don't notice it. From a phenomenological point of view, this isn't very surprising: Our impression of continuity is based on short-time memory and intentionality, not on the permanence and continuity of our subjective experience itself. Even if we had a single, isolated mental experience, it would have the same quality and provide the same sensation of time as a continuous one – for as the poet says, eternity lies in every moment... The continuity of time only applies at the material level, while the mental world only picks out single "snapshots" of the state of the material world, so to say. Determining when these mental experiences take place cannot be achieved by investigating their subjective content alone; only the elusive effects of the simultaneous collapse of the quantum state could indicate this. So for the sake of simplicity, I just assumed a periodic collapse with a given elementary period, in order to have a well-defined model, even if we don't have yet any experimental clues in this respect. Of course, the collapse of the quantum state is not a local process in the sense of Relativity Theory, but Einstein-Podolsky-Rosen experiments have already shown very clearly that this non-locality is really part of Nature. And after all, who would expect mental phenomena to be local? They are not bounded to their material substrate; they don't live in the frame of space, but in another dimension of the physical world, so to say.

In the end, the model I'm proposing can be roughly described in very simple terms: A mental state is being experienced while the quantum state is undergoing an elementary unitary evolution, then a new mental state is being randomly moved to as the quantum state gets projected to the corresponding subspace, an so on. In the meanwhile, this almost sounds trivial to me, so I guess I'm eventually understanding Quantum Physics, at least in this form. This alone would be a revolution in this field of science. But I'm not interested in pretending to have discovered deep truths about "the inmost force which binds the world", to speak with Goethe; I just wanted to show that it is possible, and actually quite easy, to give Quantum Field Theory a form and an interpretation which make it a formally and conceptually closed theory, capable of giving a well-defined answer to any question we can ask it – even if we may eventually find out that it wasn't the right one. This interpretation challenges all existing ones insofar as it is the first time that this degree of conceptual precision and formal well-definedness has been reached, and I hope this will be motivation enough for others to work out alternative interpretations and achieve the same level of quality – so that we can finally know what Quantum Theory is actually about...

Introduction

Interpreting Quantum Field Theory

The birth of Quantum Physics in the 1920s has been marked by a long period of intense controversies about its interpretation, which has been recently reviewed by Juan Miguel Marin in his paper *'Mysticism' in quantum mechanics: the forgotten controversy* [15]. The Copenhagen interpretation which emerged from these debates is dominating the scene since the 1950s-1960s, certainly not because it is intrinsically better than others, but because it seems to challenge the materialist world view of classical, 'everyday' physics as little as possible. Since the 1970s, however, numerous alternative interpretations have been proposed and further developed: Pilot-Wave Theory, Dynamical Collapse Theories, Many-Worlds interpretation, Many-Minds interpretation, Decoherent Histories... All these attempts to give Quantum Physics a sound interpretation are facing the problem that the mathematical theory itself, in the form of the Standard Model of Quantum Field Theory (or of slight variants regarding the existence of the Higgs field, of neutrino masses...), is still ill-defined, and that it is therefore impossible to assign a physical or metaphysical meaning to the fundamental mathematical entities of the theory, *i.e.* to define an ontology. Of course, it is possible to choose a specific, mathematically well-defined regularization of the theory for this purpose, but since renormalization methods are leading, in the singular limit of the original theory, to the same results for different regularization schemes, we don't know which regularization is the right one, and as a consequence we don't know either which are the fundamental mathematical entities to be interpreted. Quite surprisingly, however, it seems that this issue has never been seriously accounted for yet. Existing interpretations of Quantum Physics are either restricted to non-relativistic Quantum Mechanics, which is no fundamental theory, or are formulated so vaguely that they are hardly more than the mere idea of an interpretation, which has made many physicists doubt that such an interpretation is possible at all. This book will have reached his goal if it convinces the reader of the contrary and helps interpretation issues recovering again their place at the heart of the research on Quantum Field Theory. For this purpose, I shall take as an example a lattice regularization scheme, formulate in this well-defined framework a rather conservative interpretation inspired by Spinoza's philosophy and show that classical philosophical questions can be formulated as simple physical hypotheses in the frame of the resulting naturalistic metaphysics.

Philosophical motivation

Most philosophers have ascribed a central role to the ethics in their work, as the answer to the question "How should I live?" requires a preliminary reflection about all the fields of our existence, from metaphysics via physics, psychology and morals up to politics. Up to the emergence of Quantum Field Theory in the late 1920s, philosophers have always been able to integrate the knowledge gathered in the field of physics into their world view: In 17th-century Europe, for instance, the Dutch philosopher Baruch Spinoza worked out in his *Ethics* [18] the deterministic materialism of classical mechanics and based his philosophy on the idea that everything in Nature happens according to the divine necessity, both at the material and at the spiritual level. In fact, from the Antics up to the Age of Enlightenment, physicists used to consider themselves primarily as Nature philosophers. In the modern ages, however, the scientific community began to split under the influence of industrial work organization into small groups of specialists lacking interdisciplinary skills. Nowadays, mainstream physicists even consider philosophical interpretations of Physics as non-scientific and pointless. Needless to say, such a lack of intellectual rigor has had serious consequences for the conceptual and formal quality of physical theories. In the case of Quantum Field Theory, this attitude has resulted in the fact that, for the last eighty years, no consensus could be reached on its two major issues, known as the *measurement problem* and the *main issue*. The latter is a formal issue consisting in conceiving a mathematically well-defined quantum field theory formally compatible to Special Relativity*, which would be highly desirable but is thought to be technically impossible, although this hasn't been proved definitely yet. The former is an interpretation issue concerning the relation between "mind", *i.e.* the primitive form of our experience of the world, and "body", *i.e.* the material world described in terms of quantum fields. There have been numerous propositions for this interpretation, from the very beginning of Quantum Field Theory in the 1920s and 1930s until now, but as long as the mathematical formalism of the theory is ill-defined, these interpretations cannot be formulated precisely either. Strictly speaking, however, the whole theory doesn't make any sense if this interpretation issue doesn't become precisely answered. In order to give a sound philosophical interpretation of Quantum Field Theory, it is therefore necessary in the first place to put the mathematical formalism on a well-defined basis. The theory cannot be formally compatible with Special Relativity if the main issue really cannot be solved; Relativity Theory should then emerge as an approximation at usual energy and distance scales. In this book, I shall propose an answer to both issues, describe very precisely the form of the relation between mind and body according to Quantum Physics and thus lay the foundations of an ethic taking into account the world view sustained by Quantum Field Theory.

---

*Precisely, one requires that the classical Lagrangian used in the heuristical construction of the theory be Poincaré invariant and describes local (contact) interactions of point particles in the Minkowski space-time.

Abstract

In the formulation of Quantum Field Theory proposed in this book, Nature presents the two aspects of a material and of a mental world in mutual interaction. The mental world can be adequately experienced by the collectivity of all sentient beings: A state of the mental world is given by the number of subjects having each possible subjective experience. Though we are experiencing this mental world directly, we only experience it partially under the aspect of a single subjective experience, and must communicate with others subjects in order to get closer to an adequate representation of the mental world as a whole. However, communication happens only via the material world, which is an aspect of Nature that we don't experience directly but only through its influence on our subjective experience. This material world, best described in terms of quantum fields, is by nature holistic and doesn't involve precise boundaries of individual bodies. A state of this material world is given by a quantum superposition of so-called localized states, which are given by the number of elementary particles of each kind present at each point of space. Each quantum state can be uniquely decomposed into a sum of components corresponding to each possible mental state, and this decomposition defines a probability law on the set of all possible mental states. The joined temporal evolution of both aspects of Nature is a tree-steps process repeated indefinitely: First, the initial state of the material world undergoes a deterministic, Hamiltonian evolution of a given, "elementary" duration. Then, the final quantum state defines a probability law according to which a mental state is being selected and becomes experienced by a various number of subjects. Finally, the component of the quantum state corresponding to the selected mental state becomes the initial state of the next evolution process. In this world view, the mystery of consciousness consists in the fact that there is, to some extent, an adequation between subjective experiences and the physical processes happening in the corresponding quantum states, e.g. the biological processes of consciousness within a human brain.

Overview

This book begins in chapter 1 with the formulation of a mathematically well-defined frame for any theory of mutually interacting quantum fields of point particles. Well-definedness is achieved by making sure that the Hamilton space of the quantum states is finite dimensional, so that the Hamiltonian evolution is trivially well-defined for any interaction Hamiltonian. The "ingredients" of this mathematical frame are already well-known in Quantum Field Theory: Space is supposed to have the structure of a finite three dimensional lattice, the definition of the kinetic energy Hamiltonian making use of the SLAC derivative, and the occupation number of single modes of the particle fields is supposed to be bounded for bosons as well as for fermions.

The Hamiltonian evolution of quantum fields is then defined very classically for an arbitrary interaction Hamiltonian in chapter 2, where general results of scattering theory are being derived.

A general model for the mental world is defined in chapter 3 and the joint stochastic evolution of the material and mental worlds in chapter 4. The basic idea of this model – that "mind causes collapse" – isn't quite new, as it has first been formulated by John von Neumann [16] and was once known as the 'standard interpretation' of Quantum Mechanics. As far as I know, however, it is the first time with this book that a precise interpretation of a mathematically well-defined Quantum Field Theory has ever been given. This provides thus the first sound basis for a discussion of the philosophical implications of the theory, which is the main goal of this book.

The metaphysics of the theory are been sketched in chapter 5 and its interpretation discussed at length in chapter 6. A few classical philosophical questions are then addressed on this background in chapter 7.

The interaction Hamiltonian of Quantum Electrodynamics is then defined in chapter 8 and, as an example, the semi-classical cross-section of Coulomb scattering is calculated to the leading order in chapter 10.

Finally, some usual mathematical functions, notations and operators are being defined in the appendix.

Part I

Material world

Chapter 1

Quantum fields

> The first simplification to be considered involves the very existence of the theory.
>
> John Collins,
> *Renormalization* [5]

The aim of this chapter is to develop a well-defined, divergence-free mathematical formalism for the Standard Model of particle physics. To achieve this, we suppose that elementary particles are bounded to a finite lattice, also a finite set of world lines in the flat space-time (so that the particle field only has a finite number of modes), and that there is a maximum occupation number for any single mode of the field, for bosons as well as for fermions. This makes the Hilbert space of the states of the universe finite dimensional, so that the theory is trivially well-defined. We will develop here a general formalism, valid for any set of elementary particles and for any form of the interaction Hamiltonian, and define the notations used in the rest of this book.

1.1 Space

DEFINITION Space is a finite set of points of the form $[\![-N, N]\!]^3$, where the physical constant N is a positive integer.

REMARKS This constant is supposed to be a "huge" integer ($\gtrsim 10^{46}$) which hasn't been measured experimentally yet. The finiteness of space is one of the conditions of the finite dimensionality of the Hilbert space of the quantum states, which will be defined in section 1.5. This is in turn a necessary condition of the well-definedness of the evolution equation 2.1 for an arbitrary Hamiltonian operator. It is therefore a theoretical necessity, which I shall assume although this fact hasn't been proved experimentally yet.

COMMENTARIES No notion of distance emerges from this definition of space. Indeed, according to the ideas developed in Einstein's vulgarization work *Relativity: The Special and General Theory* [7], we consider that distance and duration are actually no fundamental notions but have to be defined on an empirical basis. Distance and duration are measured using physical apparatus like rods or clocks, and their theoretical definition must rely on a theoretical modeling of these apparatus and of the observer making use of them. These concepts will emerge from the evolution equation 2.1 and from the expression of the Hamiltonian operator defined in sections 2.2 and 8.5. According to this expression, we will see that space has a toroidal structure, *i.e.* that opposite points on the boundary of the lattice $[-N, N]^3$ are actually nearest neighbors. This boundary is also a mere artifact, like the boundary of a world map, and doesn't represent in any way the "frontier of the universe". The physical constant a in the expression of the interaction Hamiltonian plays the role of the lattice step, *i.e.* of the distance between nearest neighbors. It is supposed to be very small ($\lesssim 10^{-20}$ m) and hasn't yet been measured experimentally either.

COMPLEMENTS We could equivalently postulate that, in the Minkowski space-time $(\mathcal{E}, g)$, defined by:

$$\begin{aligned} \mathcal{E} &:= \mathbb{R}^4 \\ g &:= diag(1, -1, -1, -1) \\ x \cdot y &:= g_{\mu\nu} x^\mu y^\nu \end{aligned}$$

elementary particles cannot occupy an arbitrary point of space but are bounded to a finite set of $(1 + 2N)^3$ world lines x_n forming in some reference frame a finite lattice of step a:

$$x_n(\tau) := \begin{pmatrix} c\tau \\ an \end{pmatrix}$$

$$n \in [-N, N]^3$$

In a reference frame moving with a velocity v relative to the lattice, the space-time coordinates of these world lines would be given (up to a translation of the origin) by:

$$x'_n(t) = \begin{pmatrix} ct \\ an_\perp + \gamma^{-1} an_\parallel - vt \end{pmatrix}$$

$$\gamma := \frac{1}{\sqrt{1 - (v/c)^2}}$$

where we use the notations $n_\parallel := (n \cdot v)v/v^2$ and $n_\perp := n - n_\parallel$.

The lattice reference frame itself as well as the physical constants N and a are free parameters of the theory. As a working hypothesis, we will assume that the lattice reference frame corresponds to a rest frame of the cosmic microwave background radiation. The relative velocity v of the sun relative to the lattice would then verify [13]:

$$v \approx 3.7 \, 10^5 \text{ m/s}$$

We will also assume that the lattice step is of the order of the Plank length:

$$a \sim \sqrt{4\pi Gh/c^3} \approx 1.4 \, 10^{-34} \text{ m}$$

and that the lattice size is of the order of the Hubble length [10]:

$$(1 + 2N)a \quad \sim \quad R_H \approx 1.3 \; 10^{26} \text{ m}$$
$$N \quad \sim \quad 4.6 \; 10^{59}$$

Incidentally, with this values, the cosmological constant of the Λ-Cold Dark Matter model of Big-Bang cosmology coincides numerically (with a relative error of only 8%) with [14]:

$$\rho_{vac} \sim 2N\frac{hc}{a} \left((1 + 2N)a \right)^{-3} \approx 5.6 \; 10^{-10} \text{ J/m}^3$$

Deriving such a relation, however, isn't the goal of this book.

1.2 One particle states

PARTICLE TYPES We don't make in this chapter any assumption about the existing particle types, e.g. electrons, positrons and photons. We are noting them ϕ in a generic way. The corresponding spin states λ depend implicitly, in the notations, on the particle type. The spin state influences the way a particle interacts with other particle fields; this effect is described quantitatively in the expression of the spinor operators in appendix C.1 and C.3.

DEFINITION The (hypothetical) quantum state $|\Psi\rangle$ in which the universe only contains a single particle, of type ϕ, at point $\boldsymbol{n}$ and in the spin state λ, is written:

$$|\Psi\rangle = \left| 1^{\phi}_{\boldsymbol{n},\lambda} \right\rangle$$

We postulate that a one particle state is given by any linear combination of the form:

$$|\Psi\rangle = \sum_{\phi,\boldsymbol{n},\lambda} \Psi(1^{\phi}_{\boldsymbol{n},\lambda}) \left| 1^{\phi}_{\boldsymbol{n},\lambda} \right\rangle$$

with arbitrary complex coefficients $\Psi(1^{\phi}_{\boldsymbol{n},\lambda})$. The set of all these vectors, taken as an orthonormal basis, forms a finite dimensional Hilbert space written $\mathcal{H}_1$ and given by:

$$\mathcal{H}_1 := \bigoplus_{\phi,\boldsymbol{n},\lambda}^{\perp} \mathbb{C} \left| 1^{\phi}_{\boldsymbol{n},\lambda} \right\rangle$$

MOMENTUM REPRESENTATION We postulate that the momentum $\boldsymbol{p}$ of a particle in the lattice reference frame can only take values of the form:

$$\boldsymbol{p} = \frac{h}{a}\boldsymbol{q}$$
$$\boldsymbol{q} \in \left(\frac{[\![-N, N]\!]}{1 + 2N} \right)^3$$

and that the (hypothetical) quantum state in which the universe only contains a single particle, of type ϕ, in the spin state λ with the momentum $h\boldsymbol{q}/a$ in the lattice reference frame, is given by:

$$\left|1_{\boldsymbol{q},\lambda}^{\phi}\right\rangle := (1+2\mathrm{N})^{-3/2} \sum_{\boldsymbol{n}} \exp\left(\mathrm{i}2\pi\boldsymbol{n}\cdot\boldsymbol{q}\right) \left|1_{\boldsymbol{n},\lambda}^{\phi}\right\rangle$$

These vectors form an orthonormal basis of the Hilbert space $\mathcal{H}_1$ and we will use the notation:

$$|\Psi\rangle = \sum_{\phi,\boldsymbol{q},\lambda} \tilde{\Psi}(1_{\boldsymbol{q},\lambda}^{\phi}) \left|1_{\boldsymbol{q},\lambda}^{\phi}\right\rangle$$

NOTATION In order to simplify the notations, when defining and using periodical functions on all $\boldsymbol{q} \in \mathbb{R}^3$, we will define $\underline{\boldsymbol{q}} \in \left]-\frac{1}{2},\frac{1}{2}\right]^3$ by the equivalence relation $\underline{\boldsymbol{q}}-\boldsymbol{q} \in \mathbb{Z}^3$. We have then in particular $\underline{\boldsymbol{q}} = \boldsymbol{q}$ for all $\boldsymbol{q} \in \left(\frac{[\![-\mathrm{N},\mathrm{N}]\!]}{1+2\mathrm{N}}\right)^3$ and $\underline{\boldsymbol{q}} \in \left(\frac{[\![-\mathrm{N},\mathrm{N}]\!]}{1+2\mathrm{N}}\right)^3$ for all $\boldsymbol{q} \in \left(\frac{\mathbb{Z}}{1+2\mathrm{N}}\right)^3$.

1.3 Position and momentum operators

DEFINITION In the lattice reference frame, we define on $\mathcal{H}_1$ the position and momentum operators by:

$$\widehat{\boldsymbol{r}} \left|1_{\boldsymbol{n},\lambda}^{\phi}\right\rangle := \mathrm{a}\boldsymbol{n} \left|1_{\boldsymbol{n},\lambda}^{\phi}\right\rangle$$

$$\widehat{\boldsymbol{p}} \left|1_{\boldsymbol{q},\lambda}^{\phi}\right\rangle := \frac{\mathrm{h}}{\mathrm{a}}\boldsymbol{q} \left|1_{\boldsymbol{q},\lambda}^{\phi}\right\rangle$$

REMARK This definition of the momentum operator follows the same principle as the SLAC derivative [17], but can be expressed as a proper eigenvalue equation, since momentum eigenstates are well-defined on a finite lattice.

COMPLEMENTS In another reference frame, moving with a velocity $\boldsymbol{v}$ relative to the lattice, these operators are given (up to a translation of the origin) by:

$$\widehat{\boldsymbol{r}} \left|1_{\boldsymbol{n},\lambda}^{\phi}\right\rangle := \left(\mathrm{a}\boldsymbol{n}_\perp + \gamma^{-1}\mathrm{a}\boldsymbol{n}_\parallel - \boldsymbol{v}t\right) \left|1_{\boldsymbol{n},\lambda}^{\phi}\right\rangle$$

$$\widehat{\boldsymbol{p}} \left|1_{\boldsymbol{q},\lambda}^{\phi}\right\rangle := \left(\frac{\mathrm{h}}{\mathrm{a}}\boldsymbol{q}_\perp + \gamma\frac{\mathrm{h}}{\mathrm{a}}\boldsymbol{q}_\parallel - \gamma\frac{E_{\boldsymbol{q}}^{\phi}}{\mathrm{c}^2}\boldsymbol{v}\right) \left|1_{\boldsymbol{q},\lambda}^{\phi}\right\rangle$$

where $E_{\boldsymbol{q}}^{\phi}$ is the kinetic energy of the particle in the lattice reference frame, defined as a function of its (bare) rest mass m_ϕ by:

$$E_{\boldsymbol{q}}^{\phi} := \sqrt{\left(\mathrm{m}_\phi\mathrm{c}^2\right)^2 + \left(\frac{\mathrm{hc}}{\mathrm{a}}\boldsymbol{q}\right)^2}$$

Similarly, we define the relativistic factors $\beta_{\boldsymbol{q}}^{\phi}$ and $\gamma_{\boldsymbol{q}}^{\phi}$, with the help of the reduced mass $\mathrm{M}_\phi := \mathrm{m}_\phi\mathrm{ac/h}$, by:

$$\beta_{\boldsymbol{q}}^{\phi} := \frac{\boldsymbol{q}}{\sqrt{\mathrm{M}_\phi^2 + \underline{\boldsymbol{q}}^2}}$$

$$\gamma_q^\phi := \sqrt{1 + \left(\frac{q}{M_\phi}\right)^2}$$

and the velocity by $v_q^\phi := \beta_q^\phi c$.

1.4 Wave function

COMPLEMENTS We can associate following wave function components to each one particle state:

$$\Psi_\lambda^\phi (x) := (1 + 2N)^{-3/2} \sum_q \widetilde{\Psi}(1_{q,\lambda}^\phi) \exp\left(i2\pi \frac{x \cdot q}{a}\right)$$

Eigenstates of the momentum operator are thus associated with plane waves on $\mathbb{R}^3$. Equivalently, we can write:

$$\Psi_\lambda^\phi (x) = \sum_n \Psi(1_{n,\lambda}^\phi)\delta_x (x - an)$$

$$\delta_x (x) := (1 + 2N)^{-3} \prod_i \frac{\sin(\pi x_i/a)}{\sin(\pi x_i/(1 + 2N)a)}$$

We define thus an isomorphism between a finite set, indexed on (ϕ, λ), of complementary subspaces of $\mathcal{H}_1$, and a finite dimensional subspace of $C^\infty (\mathbb{R}^3, \mathbb{C})$ containing functions of period $(1 + 2N)a$ along each coordinate.

In that space, the (image of the) momentum operator acts according to:

$$\widehat{p}\Psi_\lambda^\phi (x) = \frac{h}{i2\pi}\nabla\Psi_\lambda^\phi (x)$$

The dynamic of the free fields on the lattice is also identical to the usual dynamic of the free fields on the continuum in the box $]-(N + \frac{1}{2})a, (N + \frac{1}{2})a[^3$ with periodical boundary conditions.

1.5 Many particles states

The quantum state $|\Psi\rangle$ in which each point n is being occupied by $N_{n,\lambda}^\phi$ particles of each type ϕ in each spin state λ is written:

$$|\Psi\rangle = \left|(N_{n,\lambda}^\phi)\right\rangle$$

and is called a "localized state". We postulate that a many particles state is given by any linear combination of the form:

$$|\Psi\rangle = \sum_{(N_{n,\lambda}^\phi)} \Psi\left((N_{n,\lambda}^\phi)\right) \left|(N_{n,\lambda}^\phi)\right\rangle$$

$$N_{n,\lambda}^\phi \in [\![0, N_\phi^{max}]\!]$$

where the (finite) integer N_ϕ^{max} is the maximum occupation number of the field ϕ.

The set of all these vectors, taken as an orthonormal basis, forms a finite dimensional Hilbert space given by:

$$\mathcal{H} := \overset{\perp}{\underset{(N_{n,\lambda}^\phi)}{\bigoplus}} \mathbb{C} \left|(N_{n,\lambda}^\phi)\right\rangle$$

and the basis of the localized states is called "position basis".

REMARK For fermions, we have experimentally $N_\phi^{max} = 1$. For bosons, no upper limit of the occupation number is experimentally known; a lower limit of about $N_\gamma^{max} \gtrsim 10^{21}$ for photons has been reached experimentally by high intensity lasers.

1.6 Creation and annihilation operators

DEFINITION The annihilation operators are given by:

$$\widehat{a^\phi}_{n,\lambda} \left|(N_{n,\lambda}^\phi)\right\rangle := \begin{cases} \left|(N_{n,\lambda}^\phi) - 1_{n,\lambda}^\phi\right\rangle & \text{if } N_{n,\lambda}^\phi > 0 \\ 0 & \text{otherwise} \end{cases}$$

and the creation operators by:

$$\widehat{a^\phi}^\dagger_{n,\lambda} \left|(N_{n,\lambda}^\phi)\right\rangle := \begin{cases} \left|(N_{n,\lambda}^\phi) + 1_{n,\lambda}^\phi\right\rangle & \text{if } N_{n,\lambda}^\phi < N_\phi^{max} \\ 0 & \text{otherwise} \end{cases}$$

The (hypothetical) state of the universe in which no particles are present is written:

$$|\Psi\rangle = |\Omega\rangle := \left|(0_{n,\lambda}^\phi)\right\rangle$$

REMARK The annihilation (resp. creation) operators form a (finite) set of generators of a commutative algebra $\mathcal{A}$ (resp. $\mathcal{A}^\dagger$). Any state of the universe can be obtained by applying creation operators on the vacuum according to:

$$|\Psi\rangle = \widehat{\Psi}^\dagger |\Omega\rangle$$

$$\widehat{\Psi}^\dagger := \sum_{(N_{n,\lambda}^\phi)} \Psi\left((N_{n,\lambda}^\phi)\right) \prod_{\phi,n,\lambda} \left(\widehat{a^\phi}^\dagger_{n,\lambda}\right)^{N_{n,\lambda}^\phi}$$

associating thus an operator $\widehat{\Psi}^\dagger \in \mathcal{A}^\dagger$ to each vector $|\Psi\rangle \in \mathcal{H}$ canonically.

COMMENTARIES We are defining here at purpose very basic creation and annihilation operators. The normalization factor relevant for boson fields and the antisymmetry factor relevant for fermion fields are included explicitly in the interaction Hamiltonian, e.g. in the photon spinor operators defined in appendix C.1 and in the Dirac spinor operators defined in appendix C.3.

1.7 Plane wave field modes

DEFINITION Creation and annihilation operators can also be defined for the plane wave modes of the field by:

$$\widehat{a^{\phi}}_{\boldsymbol{q},\lambda} \;:=\; (1+2\mathrm{N})^{-3/2} \sum_{\boldsymbol{n}} \exp\left(-\mathrm{i}2\pi \boldsymbol{n}\cdot\boldsymbol{q}\right)\widehat{a^{\phi}}_{\boldsymbol{n},\lambda}$$

$$\widehat{a^{\phi}}^{\dagger}_{\boldsymbol{q},\lambda} \;:=\; (1+2\mathrm{N})^{-3/2} \sum_{\boldsymbol{n}} \exp\left(\mathrm{i}2\pi \boldsymbol{n}\cdot\boldsymbol{q}\right)\widehat{a^{\phi}}^{\dagger}_{\boldsymbol{n},\lambda}$$

Note that this definition can be extended to all $\boldsymbol{q}\in\mathbb{R}^3$. The plane wave states of the field are then defined by:

$$\left|(N^{\phi}_{\boldsymbol{q},\lambda})\right\rangle := \prod_{\phi,\boldsymbol{q},\lambda}\left(\widehat{a^{\phi}}^{\dagger}_{\boldsymbol{q},\lambda}\right)^{N^{\phi}_{\boldsymbol{q},\lambda}}|\Omega\rangle$$

These vectors form an orthonormal basis of the Hilbert space $\mathcal{H}$ called the "momentum basis" and we will use the notation:

$$|\Psi\rangle = \sum_{(N^{\phi}_{\boldsymbol{q},\lambda})}\widetilde{\Psi}\left((N^{\phi}_{\boldsymbol{q},\lambda})\right)\left|(N^{\phi}_{\boldsymbol{q},\lambda})\right\rangle$$

REMARK The decomposition of the plane wave state $\left|(N'^{\phi}_{\boldsymbol{q},\lambda})\right\rangle$ on the position basis is given by:

$$\left\langle (N^{\phi}_{\boldsymbol{n},\lambda})|(N'^{\phi}_{\boldsymbol{q},\lambda})\right\rangle = \left[\prod_{\phi,\lambda}\delta\left(N'^{\phi}_{\lambda}-N^{\phi}_{\lambda}\right)\right]\psi\left((\boldsymbol{q}^{\phi,\lambda}_j),(\boldsymbol{n}^{\phi,\lambda}_j)\right)$$

$$\psi\left((\boldsymbol{q}^{\phi,\lambda}_j),(\boldsymbol{n}^{\phi,\lambda}_j)\right) := \prod_{\substack{\phi,\lambda\\ N^{\phi}_{\lambda}\neq 0}}\frac{(1+2\mathrm{N})^{-3N^{\phi}_{\lambda}/2}}{\prod_{\boldsymbol{n}}N^{\phi}_{\boldsymbol{n},\lambda}!}\sum_{\sigma\in\mathfrak{S}_{N^{\phi}_{\lambda}}}\prod_{j=1}^{N^{\phi}_{\lambda}}\exp\left(\mathrm{i}2\pi\boldsymbol{n}^{\phi,\lambda}_{\sigma_j}\cdot\boldsymbol{q}^{\phi,\lambda}_j\right)$$

where we use the notations $N'^{\phi}_{\lambda} := \sum_{\boldsymbol{q}}N'^{\phi}_{\boldsymbol{q},\lambda}$ and $N^{\phi}_{\lambda} := \sum_{\boldsymbol{n}}N^{\phi}_{\boldsymbol{n},\lambda}$, where $\mathfrak{S}_{N^{\phi}_{\lambda}}$ denotes the symmetric group of order N^{ϕ}_{λ} and where we have chosen for each mode (ϕ,λ) of the field the families $(\boldsymbol{n}^{\phi,\lambda}_j)$ and $(\boldsymbol{q}^{\phi,\lambda}_j)$ so that:

$$\left|(N^{\phi}_{\boldsymbol{n},\lambda})\right\rangle = \prod_{\phi,\lambda,j}\widehat{a^{\phi}}^{\dagger}_{\boldsymbol{n}^{\phi,\lambda}_j,\lambda}|\Omega\rangle$$

$$\left|(N'^{\phi}_{\boldsymbol{q},\lambda})\right\rangle = \prod_{\phi,\lambda,j}\widehat{a^{\phi}}^{\dagger}_{\boldsymbol{q}^{\phi,\lambda}_j,\lambda}|\Omega\rangle$$

In the definition of $\psi\left((\boldsymbol{q}^{\phi,\lambda}_j),(\boldsymbol{n}^{\phi,\lambda}_j)\right)$, we used for convenience the symbols $N^{\phi}_{\boldsymbol{n},\lambda}$ and N^{ϕ}_{λ}, which can be defined as a function of $(\boldsymbol{n}^{\phi,\lambda}_j)$ with $N^{\phi}_{\boldsymbol{n},\lambda} := \left|\{j \mid \boldsymbol{n}^{\phi,\lambda}_j = \boldsymbol{n}\}\right|$.

1.8 Particle number operators

DEFINITION The particle number operators are defined by:

$$\widehat{N^\phi}_{n,\lambda} \left| (N^\phi_{n,\lambda}) \right\rangle \; := \; N^\phi_{n,\lambda} \left| (N^\phi_{n,\lambda}) \right\rangle$$

$$\widehat{N^\phi}_{q,\lambda} \left| (N^\phi_{q,\lambda}) \right\rangle \; := \; N^\phi_{q,\lambda} \left| (N^\phi_{q,\lambda}) \right\rangle$$

The total particle number operator is defined as the (finite) sum:

$$\hat{N} := \sum_{\phi,n,\lambda} \widehat{N^\phi}_{n,\lambda} = \sum_{\phi,q,\lambda} \widehat{N^\phi}_{q,\lambda}$$

Its eigenspace for the eigenvalue N is written $\mathcal{H}_N$ and its elements are called "N particle states" of the field. The Hilbert space can be decomposed into a (finite) sum of the form:

$$\mathcal{H} = \overset{\perp}{\underset{N}{\bigoplus}} \mathcal{H}_N$$

The maximum number of particles in a N particle state is given by $N = (1 + 2N)^3 \sum_{\phi,\lambda} N^{max}_\phi$.

Chapter 2

Hamiltonian evolution

2.1 Schrödinger equation

We postulate that the state of the quantum field evolves according to an equation of the form:

$$\frac{\mathrm{d}}{\mathrm{d}t}\,|\Psi\rangle = -\mathrm{i}2\pi\frac{1}{\mathrm{h}}\widehat{\mathrm{H}}\,|\Psi\rangle$$

called "Schrödinger equation" where $\widehat{\mathrm{H}}$ is the (time independent) Hamiltonian operator of the field. This operator is supposed to be hermitian and is therefore diagonalizable (with real eigenvalues) on the finite dimensional Hilbert space $\mathcal{H}$. The equation can also be integrated as:

$$
\begin{aligned}
|\Psi(t)\rangle &= \widehat{\mathrm{U}}(t,t_0)\,|\Psi(t_0)\rangle \\
\widehat{\mathrm{U}}(t,t_0) &:= \exp\left(-\mathrm{i}2\pi\frac{t-t_0}{\mathrm{h}}\widehat{\mathrm{H}}\right)
\end{aligned}
$$

2.2 Kinetic energy Hamiltonian

The Hamiltonian operator of the field can be separated into a kinetic energy Hamiltonian depending only on the momentum of the particles and an interaction term as follows:

$$\widehat{\mathrm{H}} = \widehat{\mathrm{H}}_0 + \widehat{\mathrm{H}}'$$

In the lattice reference frame, the kinetic energy Hamiltonian is given by:

$$\widehat{\mathrm{H}}_0 := \sum_{\phi,\boldsymbol{q},\lambda} E_{\boldsymbol{q}}^{\phi}\widehat{N^{\phi}}_{\boldsymbol{q},\lambda}$$

In another reference frame, moving with a velocity $\boldsymbol{v}$ relative to the lattice, this operator is given by:

$$\widehat{\mathrm{H}}_0 := \sum_{\phi,\boldsymbol{q},\lambda} \gamma\left\{E_{\boldsymbol{q}}^{\phi} - \frac{\mathrm{h}}{\mathrm{a}}\boldsymbol{q}\cdot\boldsymbol{v}\right\}\widehat{N^{\phi}}_{\boldsymbol{q},\lambda}$$

2.3 Interaction picture

The kinetic energy Hamiltonian can be integrated as:

$$\widehat{U}_0(t, t_0) := \exp\left(-i2\pi\frac{t - t_0}{h}\widehat{H}_0\right)$$

The state of the quantum field in the interaction picture is defined in such a way that it would be a time constant if the interaction term $\widehat{H}'$ vanishes:

$$|\Psi_I\rangle := \widehat{U}_0(0, t)\ |\Psi\rangle$$

The Hamiltonian operator in the interaction picture is defined is such a way that the state of the quantum field in the interaction picture obeys following Schrödinger-like equation, where the Hamiltonian is time-dependent:

$$\frac{d}{dt}\ |\Psi_I\rangle\ =\ -i2\pi\frac{1}{h}\widehat{H}_I\ |\Psi_I\rangle$$
$$\widehat{H}_I\ :=\ \widehat{U}_0(0, t)\widehat{H}'\widehat{U}_0(t, 0)$$

The integration of this equation yields to:

$$|\Psi_I(t)\rangle = \widehat{U}_I(t, t_0)\ |\Psi_I(t_0)\rangle$$

where the evolution operator in the interaction picture is given by a series of the form (assuming $t > t_0$):

$$\widehat{U}_I(t, t_0)\ :=\ \mathbb{1} + \sum_{n=1}^{\infty}\widehat{U}_I^{(n)}(t, t_0)$$
$$\widehat{U}_I^{(n)}(t, t_0)\ :=\ \left(\frac{-i2\pi}{h}\right)^n\int_{t>t_n>\cdots>t_1>t_0} dt_1\cdots dt_n$$
$$\widehat{U}_0(0, t_n)\widehat{H}'\widehat{U}_0(t_n, t_{n-1})\cdots\widehat{U}_0(t_2, t_1)\widehat{H}'\widehat{U}_0(t_1, 0)$$

The evolution operator in the interaction picture verifies:

$$\widehat{U}_I(t, t_0) = \widehat{U}_0(0, t)\widehat{U}(t, t_0)\widehat{U}_0(t_0, 0)$$

The usual evolution operator can also be written too as a series of the form:

$$\widehat{U}(t, t_0)\ :=\ \sum_{n=0}^{\infty}\widehat{U}^{(n)}(t, t_0)$$
$$\widehat{U}^{(0)}(t, t_0)\ :=\ \widehat{U}_0(t, t_0)$$
$$\widehat{U}^{(n)}(t, t_0)\ :=\ \left(\frac{-i2\pi}{h}\right)^n\int_{t>t_n>\cdots>t_1>t_0} dt_1\cdots dt_n$$
$$\widehat{U}_0(t, t_n)\widehat{H}'\widehat{U}_0(t_n, t_{n-1})\cdots\widehat{U}_0(t_2, t_1)\widehat{H}'\widehat{U}_0(t_1, t_0)$$

2.4   Transition amplitudes

In scattering experiments, the evolution operator in the interaction picture is often called "scattering operator". In this context, cross sections are usually calculated in the limit $t_0 \to -\infty$ and $t \to +\infty$, so the variables t_0 and t are implicit in the notation:

$$\widehat{S} := \widehat{U}_I(t, t_0)$$

Its matrix elements, called "scattering amplitudes" and written:

$$
\begin{aligned}
S_{fi} &:= \langle \Psi_f | \, \widehat{S} \, | \Psi_i \rangle \\
&= \langle \Psi_f | \, \widehat{U}_I(t, t_0) \, | \Psi_i \rangle
\end{aligned}
$$

can be developed in a series of the form (assuming $t > t_0$):

$$
\begin{aligned}
S_{fi} &= \sum_{n=0}^{\infty} S_{fi}^{(n)} \\
S_{fi}^{(0)} &:= \langle \Psi_f | \Psi_i \rangle \\
S_{fi}^{(n)} &:= \left(\frac{-i2\pi}{h} \right)^n \int_{t > t_n > \cdots > t_1 > t_0} dt_1 \cdots dt_n \\
&\quad \langle \Psi_f | \, \widehat{U}_0(0, t_n) \widehat{H}' \widehat{U}_0(t_n, t_{n-1}) \cdots \widehat{U}_0(t_2, t_1) \widehat{H}' \widehat{U}_0(t_1, 0) \, | \Psi_i \rangle
\end{aligned}
$$

For plane wave states $|\Psi_i\rangle = \left| (N_{i\boldsymbol{q},\lambda}^{\phi}) \right\rangle$ and $|\Psi_f\rangle = \left| (N_{f\boldsymbol{q},\lambda}^{\phi}) \right\rangle$, they are directly related to the matrix elements of the evolution operator, called "transition amplitudes", by:

$$
\begin{aligned}
U_{fi}(t, t_0) &= \exp\left(-i2\pi(tE_f - t_0 E_i)/h \right) S_{fi} \\
U_{fi}(t, t_0) &:= \langle \Psi_f | \, \widehat{U}(t, t_0) \, | \Psi_i \rangle \\
E_i &:= \langle \Psi_i | \, \widehat{H}_0 \, | \Psi_i \rangle \\
E_f &:= \langle \Psi_f | \, \widehat{H}_0 \, | \Psi_f \rangle
\end{aligned}
$$

The transition amplitude from a plane wave state $|\Psi_i\rangle = \left| (N_{i\boldsymbol{q},\lambda}^{\phi}) \right\rangle$ to a localized state $|\Psi_f\rangle = \left| (N_{f\boldsymbol{n},\lambda}^{\phi}) \right\rangle$ can in turn be written as:

$$
U_{fi}(t, t_0) = \sum_{(N_{f\boldsymbol{q},\lambda}^{\phi})} S_{fi} \exp\left(i2\pi t_0 E_i/h \right) \psi\left((\boldsymbol{q}_j^{\phi,\lambda}), (\boldsymbol{n}_j^{\phi,\lambda}), t \right)
$$

with:

$$
\psi\left((\boldsymbol{q}_j^{\phi,\lambda}), (\boldsymbol{n}_j^{\phi,\lambda}), t \right) := \prod_{\substack{\phi,\lambda \\ N_{f\lambda}^{\phi} \neq 0}} \frac{(1+2N)^{-3N_{f\lambda}^{\phi}/2}}{\prod_{\boldsymbol{n}} N_{f\boldsymbol{n},\lambda}^{\phi}!} \sum_{\sigma \in \mathfrak{S}_{N_{f\lambda}^{\phi}}} \prod_{j=1}^{N_{f\lambda}^{\phi}}
$$

$$
\exp\left(i2\pi(\boldsymbol{n}_{\sigma_j}^{\phi,\lambda} \cdot \boldsymbol{q}_j^{\phi,\lambda} - E_{\boldsymbol{q}_j^{\phi,\lambda}}^{\phi} t/h) \right)
$$

where the summation runs over plane wave states $(N_{f\,\boldsymbol{q},\lambda}^{\phi})$ such that $\sum_{\boldsymbol{q}} N_{f\,\boldsymbol{q},\lambda}^{\phi} = \sum_n N_{f\,\boldsymbol{n},\lambda}^{\phi}$ for each mode (ϕ,λ) of the field, where we use the notations $S_{fi} := \left\langle (N_{f\,\boldsymbol{q},\lambda}^{\phi}) \right| \widehat{S} \left| (N_{i\,\boldsymbol{q},\lambda}^{\phi}) \right\rangle$ and $N_{f\,\lambda}^{\phi} := \sum_n N_{f\,\boldsymbol{n},\lambda}^{\phi}$, where $\mathfrak{S}_{N_{f\,\lambda}^{\phi}}$ denotes the symmetric group of order $N_{f\,\lambda}^{\phi}$ and where we have chosen for each mode (ϕ,λ) of the field the families $(\boldsymbol{n}_j^{\phi,\lambda})$ and $(\boldsymbol{q}_j^{\phi,\lambda})$ so that:

$$\left| (N_{f\,\boldsymbol{n},\lambda}^{\phi}) \right\rangle = \prod_{\phi,\lambda,j} \widehat{a^{\phi}}^{\dagger}_{\boldsymbol{n}_j^{\phi,\lambda},\lambda} \, |\Omega\rangle$$

$$\left| (N_{f\,\boldsymbol{q},\lambda}^{\phi}) \right\rangle = \prod_{\phi,\lambda,j} \widehat{a^{\phi}}^{\dagger}_{\boldsymbol{q}_j^{\phi,\lambda},\lambda} \, |\Omega\rangle$$

In the definition of $\psi\left((\boldsymbol{q}_j^{\phi,\lambda}),(\boldsymbol{n}_j^{\phi,\lambda}),t\right)$, we used for convenience the symbols $N_{f\,\boldsymbol{n},\lambda}^{\phi}$ and $N_{f\,\lambda}^{\phi}$, which can be defined as a function of $(\boldsymbol{n}_j^{\phi,\lambda})$ with $N_{f\,\boldsymbol{n},\lambda}^{\phi} := \left| \{ j \mid \boldsymbol{n}_j^{\phi,\lambda} = \boldsymbol{n} \} \right|$.

The transition amplitude from any initial state $|\Psi_i\rangle$ to a localized final state $|\Psi_f\rangle = \left| (N_{f\,\boldsymbol{n},\lambda}^{\phi}) \right\rangle$ is finally given by:

$$U_{fi}(t,t_0) = \sum_{(N_{i\,\boldsymbol{q},\lambda}^{\phi})} \sum_{(N_{f\,\boldsymbol{q},\lambda}^{\phi})} S_{fi} \widetilde{\Psi_i}\left((N_{i\,\boldsymbol{q},\lambda}^{\phi})\right) \exp\left(i2\pi t_0 E_i / h\right) \psi\left((\boldsymbol{q}_j^{\phi,\lambda}),(\boldsymbol{n}_j^{\phi,\lambda}),t\right)$$

with the same notations.

2.5 Scattering matrix

The scattering matrix can be developed quite easily on the basis of the plane wave states, *i.e.* by developing the initial and final states as:

$$|\Psi_i\rangle = \sum_{(N_{\boldsymbol{q},\lambda}^{\phi})} \widetilde{\Psi_i}\left((N_{\boldsymbol{q},\lambda}^{\phi})\right) \left| (N_{\boldsymbol{q},\lambda}^{\phi}) \right\rangle$$

$$|\Psi_f\rangle = \sum_{(N_{\boldsymbol{q},\lambda}^{\phi})} \widetilde{\Psi_f}\left((N_{\boldsymbol{q},\lambda}^{\phi})\right) \left| (N_{\boldsymbol{q},\lambda}^{\phi}) \right\rangle$$

With these notations, we have to the zeroth order:

$$S_{fi}^{(0)} = \sum_{(N_{0\,\boldsymbol{q},\lambda}^{\phi})} \overline{\widetilde{\Psi_f}\left((N_{0\,\boldsymbol{q},\lambda}^{\phi})\right)} \widetilde{\Psi_i}\left((N_{0\,\boldsymbol{q},\lambda}^{\phi})\right)$$

and to the n-th order:

$$S_{fi}^{(n)} = \sum_{(N_{k\,\boldsymbol{q},\lambda}^{\phi})}^{n} \sum_{k=0} \overline{\widetilde{\Psi_f}\left((N_{n\,\boldsymbol{q},\lambda}^{\phi})\right)} \widetilde{\Psi_i}\left((N_{0\,\boldsymbol{q},\lambda}^{\phi})\right) S_{n,\ldots,0}^{(n)}$$

$$S_{n,\ldots,0}^{(n)} := \exp\left(i2\pi \frac{t_0}{h}(E_n - E_0)\right) \left(\prod_{k=1}^{n} H'_{k,k-1}\right) S_{t-t_0}^{(n)}(E_n,\ldots,E_0)$$

$$S_{t-t_0}^{(n)}(E_n,\ldots,E_0) := \left(\frac{-\mathrm{i}2\pi}{\mathrm{h}}\right)^n \int_{t-t_0>t_n>\cdots>t_1>0}$$

$$\prod_{k=1}^{n} \exp\left(\mathrm{i}2\pi\frac{t_k}{\mathrm{h}}(E_k - E_{k-1})\right)\mathrm{d}t_n\cdots\mathrm{d}t_1$$

The functions $S_{t-t_0}^{(n)}(E_n,\ldots,E_0)$ can be calculated recursively according to:

$$S_{t-t_0}^{(1)}(E_1,E_0) = -\mathrm{i}2\pi\frac{t-t_0}{\mathrm{h}}\mathrm{esinc}\left(\frac{t-t_0}{\mathrm{h}}(E_1 - E_0)\right)$$

$$S_{t-t_0}^{(n+1)}(E_{n+1},E_n,\ldots,E_0) = \frac{1}{E_{n+1}-E_n}\left(S_{t-t_0}^{(n)}(E_{n+1},\ldots,E_0)\right.$$
$$\left.-\exp\left(\mathrm{i}2\pi\frac{t-t_0}{\mathrm{h}}(E_{n+1}-E_n)\right)S_{t-t_0}^{(n)}(E_n,\ldots,E_0)\right)$$

where the esinc function is defined as in appendix A.2. To the second order, for instance, we have:

$$S_{t-t_0}^{(2)}(E_2,E_1,E_0) = -\mathrm{i}2\pi\frac{t-t_0}{\mathrm{h}}\exp\left(\mathrm{i}\pi\frac{t-t_0}{\mathrm{h}}(E_2 - E_0)\right)\frac{1}{E_2 - E_1}$$
$$\left[\mathrm{sinc}\left(\frac{t-t_0}{\mathrm{h}}(E_2 - E_0)\right) - \exp\left(\mathrm{i}\pi\frac{t-t_0}{\mathrm{h}}(E_2 - E_1)\right)\mathrm{sinc}\left(\frac{t-t_0}{\mathrm{h}}(E_1 - E_0)\right)\right]$$

where the sinc function is defined as in appendix A.1.

2.6   Perturbative development

The explicit perturbative development of the transition amplitude between two plane wave states $|\Psi_i\rangle = \left|(N_i{}_{\boldsymbol{q},\lambda}^{\phi})\right\rangle$ and $|\Psi_f\rangle = \left|(N_f{}_{\boldsymbol{q},\lambda}^{\phi})\right\rangle$ is therefore given by:

$$\mathrm{U}_{fi}^{(0)}(t,t_0) = \exp\left(-\mathrm{i}2\pi\frac{t-t_0}{\mathrm{h}}E_f\right)\delta_{f,i}$$

$$\mathrm{U}_{fi}^{(1)}(t,t_0) = -\mathrm{i}2\pi\frac{t-t_0}{\mathrm{h}}\exp\left(-\mathrm{i}\pi\frac{t-t_0}{\mathrm{h}}(E_f + E_i)\right)\mathrm{sinc}\left(\frac{t-t_0}{\mathrm{h}}(E_f - E_i)\right)H_{f,i}'$$

$$\mathrm{U}_{fi}^{(n)}(t,t_0) = \exp\left(-\mathrm{i}2\pi\frac{t-t_0}{\mathrm{h}}E_f\right)\sum_{\substack{k=0 \\ (N_k{}_{\boldsymbol{q},\lambda}^{\phi})}}^{n} S_{t-t_0}^{(n)}(E_n,\ldots,E_0)\prod_{k=1}^{n}H_{k,k-1}'$$

where we take in the last sum $(N_0{}_{\boldsymbol{q},\lambda}^{\phi}) = (N_i{}_{\boldsymbol{q},\lambda}^{\phi})$ and $(N_n{}_{\boldsymbol{q},\lambda}^{\phi}) = (N_f{}_{\boldsymbol{q},\lambda}^{\phi})$ and where we have:

$$S_{t-t_0}^{(n)}(E_n,\ldots,E_0) := \left(\frac{-\mathrm{i}2\pi}{\mathrm{h}}\right)^n \int_{t-t_0>t_n>\cdots>t_1>0}$$

$$\prod_{k=1}^{n} \exp\left(\mathrm{i}2\pi\frac{t_k}{\mathrm{h}}(E_k - E_{k-1})\right)\mathrm{d}t_n\cdots\mathrm{d}t_1$$

More generally, the explicit perturbative development of the transition amplitude from any initial state $|\Psi_i\rangle$ to a localized final state $|\Psi_f\rangle = \left|(N_{f\,n,\lambda}^{\phi})\right\rangle$ is given by:

$$U_{fi}^{(0)}(t,t_0) = \sum_{(N_{f\,\boldsymbol{q},\lambda}^{\phi})} \widetilde{\Psi}_i\left((N_{f\,\boldsymbol{q},\lambda}^{\phi})\right) \psi\left((\boldsymbol{q}_j^{\phi,\lambda}),(\boldsymbol{n}_j^{\phi,\lambda})\right) \exp\left(-\mathrm{i}2\pi\frac{t-t_0}{\mathrm{h}}E_f\right)$$

$$U_{fi}^{(1)}(t,t_0) = -\mathrm{i}2\pi\frac{t-t_0}{\mathrm{h}} \sum_{(N_{i\,\boldsymbol{q},\lambda}^{\phi})} \widetilde{\Psi}_i\left((N_{i\,\boldsymbol{q},\lambda}^{\phi})\right) \sum_{(N_{f\,\boldsymbol{q},\lambda}^{\phi})} \psi\left((\boldsymbol{q}_j^{\phi,\lambda}),(\boldsymbol{n}_j^{\phi,\lambda})\right)$$
$$\exp\left(-\mathrm{i}\pi\frac{t-t_0}{\mathrm{h}}(E_f+E_i)\right) \mathrm{sinc}\left(\frac{t-t_0}{\mathrm{h}}(E_f-E_i)\right) H_{f,i}'$$

$$U_{fi}^{(n)}(t,t_0) = \sum_{(N_{i\,\boldsymbol{q},\lambda}^{\phi})} \widetilde{\Psi}_i\left((N_{i\,\boldsymbol{q},\lambda}^{\phi})\right) \sum_{(N_{f\,\boldsymbol{q},\lambda}^{\phi})} \psi\left((\boldsymbol{q}_j^{\phi,\lambda}),(\boldsymbol{n}_j^{\phi,\lambda})\right)$$
$$\exp\left(-\mathrm{i}2\pi\frac{t-t_0}{\mathrm{h}}E_f\right) \sum_{\substack{k=0 \\ (N_{k\,\boldsymbol{q},\lambda}^{\phi})}}^{n} \mathrm{S}_{t-t_0}^{(n)}\left(E_n,\ldots,E_0\right) \prod_{k=1}^{n} H_{k,k-1}'$$

where the summation runs over plane wave states $(N_{f\,\boldsymbol{q},\lambda}^{\phi})$ such that $\sum_{\boldsymbol{q}} N_{f\,\boldsymbol{q},\lambda}^{\phi} = \sum_{\boldsymbol{n}} N_{f\,\boldsymbol{n},\lambda}^{\phi}$ for each mode (ϕ,λ) of the field, and where we use the notation:

$$\psi\left((\boldsymbol{q}_j^{\phi,\lambda}),(\boldsymbol{n}_j^{\phi,\lambda})\right) := \prod_{\substack{\phi,\lambda \\ N_{f\,\lambda}^{\phi}\neq 0}} \frac{(1+2\mathrm{N})^{-3N_{f\,\lambda}^{\phi}/2}}{\prod_{\boldsymbol{n}} N_{f\,\boldsymbol{n},\lambda}^{\phi}!} \sum_{\sigma\in\mathfrak{S}_{N_{f\,\lambda}^{\phi}}} \prod_{j=1}^{N_{f\,\lambda}^{\phi}} \exp\left(\mathrm{i}2\pi\boldsymbol{n}_{\sigma_j}^{\phi,\lambda}\cdot\boldsymbol{q}_j^{\phi,\lambda}\right)$$

Part II

Mental world

Chapter 3

Mental states

> On the other hand I think I can safely say that nobody understands quantum mechanics.

> Richard Feynman,
> *The Character of Physical Law* [9]

3.1   The mind-body problem

Since the end of the second World War and the translation of the intellectual center of the scientific community from Europe to the United States of America, materialism, *i.e.* the complete reduction of our experience of mind to purely material processes, has become the philosophical conviction of mainstream physicists, although they still may have opposite religious beliefs as private persons. Of course, it doesn't make any doubt that the biological activity of human and similar animal brains is involved in the processing of external and internal stimuli, and it is reasonable to believe that, at the material level, conscious thinking is the emerging result of an intensive and highly parallelized information processing activity by the brain's neural network. Nevertheless, "mind", *i.e.* the form of our experience of the world, with our feelings, our body schema, memories seen with the mind's eye, melodies imagined in the mind's ear... is just not *the same* as the neural activity of an individual body (which is anyhow hardly identifiable quantum physically). Determining the relation between these two realities is the essence of the mind-body problem, which has become the most various answers over the ages. The usual divergence points arise about the questions: Do both realities exist at all, or is one of them a mere illusion? Are they independent of each other and just exist as parallel realities, or are there divergences and a mutual influence in the one, the other or both directions? In this old debate, Quantum Field Theory introduces the new idea that a mutual influence doesn't have to be a deterministic causal influence but also could be a probabilistic one, so that neither "mind" nor "body" have to be kind of a subordinated slave of its counterpart, but retain to some extent a form of "freedom" under its influence. I think this idea should have the potential to take some heat out of the debate.

3.2 Subjective experience

Each of us has a direct access to his own subjective experience and know how it "feels like" to have conscious thoughts, so I will only expose a few reflections in this place. I think that any subjective experience should be considered in its "organic" unity, that picking out single conscious thoughts and considering that the subjective experience is simply composed of these should be considered as an oversimplified and inadequate view. Within this "organic" unity, however, the intensity of consciousness may vary, focusing our awareness on some aspects rather than on others. The border between conscious and unconscious thoughts is therefore not really clear to ourselves, as there is a slow transition made up of more or less subconscious thoughts of decreasing intensity. So when I say "subjective experience", I mean in principle the unity of all conscious and subconscious thoughts, although we're not quite sure of where they end. They will, in general, contain among other things representations of a body, of its activity, of its environment, of past experiences... as well as representations of time, which make up our feeling of being continuously ourselves in the continuity of time. But I believe that this feeling of permanence of the subject is a mere illusion, for two reasons: First, this feeling is experienced in every single instant of consciousness; we could by no mean find out if we really have experienced other instants of consciousness "before" (and if there is a such thing as time in the first place) and if these instants of consciousness correspond to our current memories or not, so this feeling of permanence *could* be an illusion. In fact, if I would suddenly have the subjective experiences of another subject (with its own memories and not mine), I wouldn't even notice it! Second, subjective experience seems to cease as "our" body is dreamless sleeping, swoon or eventually die, so I think its permanence is discarded by common experience. Therefore, I don't believe that there is a such thing as a subject, or a soul, constituting a fundamental entity of the mental world, which would have an existence of its own and evolve across time, and I will only refer to *instantaneous* subjective experiences, which are not related to each other across time in the form of a personal history at a fundamental level.

3.3 Mental state

The states of the mental world are supposed to be experienced by a various number of subjects. A state $\mathfrak{M}$ of the mental world can also be described by the number $N_{\mathfrak{m}}$ of subjects having each possible subjective experience $\mathfrak{m}$. An arbitrary sequence $(N_{\mathfrak{m}})$, however, doesn't necessarily correspond to a possible mental state $\mathfrak{M}$. In fact, as a consequence of the correspondence between mental and quantum states defined subsequently and of the finite dimensionality of the Hilbert space of the quantum states, there must be a finite number of possible mental states, and a fortiori of possible subjective experiences. The set of all possible mental states is written $\mathcal{M}$.

3.4 Physical realization of mental states

The correspondence between mental and quantum states is given by a Hilbert subspace $\mathcal{H}_{\mathfrak{M}}$, called "mental subspace", associated to each possible mental state $\mathfrak{M}$ in such a way that these subspaces verify:

$$\mathcal{H} = \overset{\perp}{\underset{\mathfrak{M}}{\bigoplus}} \mathcal{H}_{\mathfrak{M}}$$

Each vector $|\Psi\rangle \in \mathcal{H}_{\mathfrak{M}} \setminus \{0\}$ is a quantum state of the universe in which the mental state $\mathfrak{M}$ is being experienced. Knowing the correspondence between mental states and mental subspaces is in essence solving the mind-body problem. As a working hypothesis, I shall assume that a mental state $\mathfrak{M} = (N_\mathfrak{m})$ is being realized physically by any quantum state describing a universe containing, for each subjective experience $\mathfrak{m}$, exactly $N_\mathfrak{m}$ human or animal brains presenting the specific activity pattern corresponding to $\mathfrak{m}$. The task of describing the possible subjective experiences belongs in principle to psychology or philosophical phenomenology, whereas the characterization of the corresponding activity patterns of the brain is the aim of cognitive neuroscience.

In mathematical terms, this hypothesis can be modeled as follows. First, the mental state $\mathfrak{M}_\Omega := (0_\mathfrak{m})$, in which no subject is having any subjective experience, is supposed to be possible, *i.e.* the corresponding subspace $\mathcal{H}_{\mathfrak{M}_\Omega}$ is supposed not to be reduced to the zero subspace. Then, for each possible subjective experience $\mathfrak{m}$, there is supposed to be a finite family of brain creation operators $(\widehat{\Psi_\mathfrak{m}^\alpha}^\dagger)$ in $\mathcal{A}^\dagger$, which are creating a single brain with an activity pattern corresponding to $\mathfrak{m}$, such that:

$$\mathcal{H}_{1_\mathfrak{m}} = \overset{\perp}{\underset{\alpha}{\bigoplus}} \widehat{\Psi_\mathfrak{m}^\alpha}^\dagger \mathcal{H}_{\mathfrak{M}_\Omega}$$

Finally, for every mental state $\mathfrak{M}$, noting $\mathfrak{M} + 1_\mathfrak{m}$ the mental state in which a single further subject is having the subjective experience $\mathfrak{m}$, the corresponding subspaces are supposed to verify:

$$\mathcal{H}_{\mathfrak{M}+1_\mathfrak{m}} = \sum_\alpha \widehat{\Psi_\mathfrak{m}^\alpha}^\dagger \mathcal{H}_{\mathfrak{M}}$$

These relations define all the subspaces $\mathcal{H}_{\mathfrak{M}}$ recursively as a function of $\mathcal{H}_{\mathfrak{M}_\Omega}$ and of the operators $\widehat{\Psi_\mathfrak{m}^\alpha}^\dagger$. If a subspace defined in this way happens to be zero (because of the existence of a maximum occupation number for single field modes), the corresponding mental state is impossible.

Given two mental states $\mathfrak{M} = (N_\mathfrak{m})$ and $\mathfrak{M}' = (N'_\mathfrak{m})$, we define the partial order relation $\mathfrak{M}' \geq \mathfrak{M}$ by $\forall \mathfrak{m}, N'_\mathfrak{m} \geq N_\mathfrak{m}$. The subspace $\mathcal{H}_{\mathfrak{M}}^+$ of the quantum states corresponding to mental states where at least $N_\mathfrak{m}$ subjects are having each subjective experience $\mathfrak{m}$ can be defined, with this notation, by:

$$\mathcal{H}_{\mathfrak{M}}^+ := \overset{\perp}{\underset{\mathfrak{M}' \geq \mathfrak{M}}{\bigoplus}} \mathcal{H}_{\mathfrak{M}'}$$

COMMENTARIES The different brain creation operators $\widehat{\Psi^{\alpha}_{m}}^{\dagger}$ corresponding to the same subjective experience m may differ for instance by a translation or a rotation of the brain, by any modification of its physical environment which doesn't involve the creation of a second brain, or by any internal modification of the quantum state of the brain itself, insofar as this doesn't influence the associated conscious thoughts. We could think for instance of neurophysiological processes involved in the unconscious brain activity or of irrelevant low-level biochemical processes.

Chapter 4

Stochastic evolution

It sounded quite a sensible voice, but it just said, "Two to the power of one hundred thousand to one against and falling," and that was all.

Douglas Adams,
The Hitchhiker's Guide to the Galaxy [1]

4.1 Collapse and mental state selection

In the joint evolution of the mental and quantum states of the universe, I suppose that the quantum state $|\Psi\rangle$ periodically becomes randomly projected into one of the mental subspaces $\mathcal{H}_{\mathfrak{M}}$, corresponding to a given mental state $\mathfrak{M}$, with a probability given by:

$$P(\mathfrak{M}) = \frac{\langle \Psi | \, \widehat{\Pi}_{\mathfrak{M}} \, | \Psi \rangle}{\langle \Psi | \Psi \rangle}$$

where $\widehat{\Pi}_{\mathfrak{M}}$ is the orthogonal projection operator on $\mathcal{H}_{\mathfrak{M}}$. Furthermore, I suppose that this projection corresponds to the fact that, in the mental world, the mental state $\mathfrak{M} = (N_{\mathrm{m}})$ is being experienced, *i.e.* that N_{m} subjects are having each possible subjective experience m. We call the material part of this process "collapse" of the quantum state of the universe and its mental counterpart "selection" of the mental state. The operators $\widehat{\Pi}_{\mathfrak{M}}$ are called "collapse operators". As a working hypothesis, we assume that the period τ of this process is of the order of the Plank time:

$$\tau \approx \sqrt{4\pi \mathrm{Gh}/\mathrm{c}^5} \approx 4.8 \; 10^{-43} \; \mathrm{s}$$

4.2 Mental evolution

Fundamentally, Quantum Field Theory also defines the probability that any given succession of mental states be experienced, an initial quantum state being given.

Explicitly, for an initial quantum state $|\Psi_i\rangle \neq 0$ at time $t_i = 0$, the probability $P_t(\mathfrak{M}_0, \ldots, \mathfrak{M}_t; |\Psi_i\rangle)$, where $t \in \mathbb{N}$, that a given sequence $\mathfrak{M}_0, \ldots, \mathfrak{M}_t$ is being experienced at times $0, \ldots, t\tau$, reads for $t = 0$:

$$P_0(\mathfrak{M}_0; |\Psi_i\rangle) = \langle \Psi_i| \, \widehat{\Pi}_{\mathfrak{M}_0} \, |\Psi_i\rangle / \langle \Psi_i|\Psi_i\rangle$$

for $t = 1$:

$$P_1(\mathfrak{M}_0, \mathfrak{M}_1; |\Psi_i\rangle) = \langle \Psi_i| \, \widehat{\Pi}_{\mathfrak{M}_0} \widehat{U}_\tau^\dagger \widehat{\Pi}_{\mathfrak{M}_1} \widehat{U}_\tau \widehat{\Pi}_{\mathfrak{M}_0} \, |\Psi_i\rangle / \langle \Psi_i|\Psi_i\rangle$$

where $\widehat{U}_\tau := \exp\left(-\mathrm{i}2\pi\widehat{H}\tau/h\right)$, and more generally for $t \geq 2$:

$$P_t(\mathfrak{M}_0, \ldots, \mathfrak{M}_t; |\Psi_i\rangle) = \langle \Psi_i| \, \widehat{\Pi}_{\mathfrak{M}_0} \widehat{U}_\tau^\dagger \widehat{\Pi}_{\mathfrak{M}_1} \cdots \widehat{U}_\tau^\dagger \widehat{\Pi}_{\mathfrak{M}_t} \widehat{U}_\tau \cdots \widehat{\Pi}_{\mathfrak{M}_1} \widehat{U}_\tau \widehat{\Pi}_{\mathfrak{M}_0} \, |\Psi_i\rangle / \langle \Psi_i|\Psi_i\rangle$$

REMARKS The probability laws obey following factorization rule, where $t' \leq t$:

$$P_t(\mathfrak{M}_0, \ldots, \mathfrak{M}_t; |\Psi_i\rangle) = P_{t'}(\mathfrak{M}_0, \ldots, \mathfrak{M}_{t'}; |\Psi_i\rangle)$$
$$P_{t-t'}(\mathfrak{M}_{t'}, \ldots, \mathfrak{M}_t; \widehat{\Pi}_{\mathfrak{M}_{t'}} \widehat{U}_\tau \cdots \widehat{\Pi}_{\mathfrak{M}_1} \widehat{U}_\tau \widehat{\Pi}_{\mathfrak{M}_0} \, |\Psi_i\rangle)$$

If the initial vector state isn't exactly known, but belongs to a given subspace $\mathcal{H}_i$, averaging on an orthonormal basis of this subspace leads to:

$$\langle P_t(\mathfrak{M}_0, \ldots, \mathfrak{M}_t; |\Psi_i\rangle)\rangle_{\mathcal{H}_i} = \mathrm{Tr}_{\mathcal{H}_i} \widehat{\Pi}_{\mathfrak{M}_0} \widehat{U}_\tau^\dagger \widehat{\Pi}_{\mathfrak{M}_1} \cdots \widehat{U}_\tau^\dagger \widehat{\Pi}_{\mathfrak{M}_t} \widehat{U}_\tau \cdots \widehat{\Pi}_{\mathfrak{M}_1} \widehat{U}_\tau \widehat{\Pi}_{\mathfrak{M}_0} / \dim \mathcal{H}_i$$

The total probability $P_t(\mathfrak{M}_t; |\Psi_i\rangle)$ that a given mental state $\mathfrak{M}_t$ is being experienced at time $t\tau$, where $t \in \mathbb{N}$, reads for $t = 0$:

$$P_0(\mathfrak{M}_0; |\Psi_i\rangle) = P_0(\mathfrak{M}_0; |\Psi_i\rangle)$$

and more generally for $t \geq 1$:

$$P_t(\mathfrak{M}_t; |\Psi_i\rangle) = \sum_{\mathfrak{M}_{t-1}} \cdots \sum_{\mathfrak{M}_0} P_t(\mathfrak{M}_0, \ldots, \mathfrak{M}_t; |\Psi_i\rangle)$$

COMPLEMENTS In the case where the actually experienced mental state has a relatively high probability, our subjective experience may give us some clues about the physics of the world we live in; on the opposite, if our mental state has a very low probability, our subjective experience has very little to do with the laws of the physical world and we live in a mere illusion of knowing something about the material reality – without having any mean of noticing it. This dilemma is very well known of particle physicists, who have to accept they cannot make more precise statements about reality than, for instance, "in the context of the standard model and in the presence of a sequential fourth family of fermions with high masses [...] a Higgs boson with mass between 144 and 207 GeV/c^2 is ruled out at 95% confidence level" [4]. Any physical model can also be conventionally, but not definitely, "ruled out" if it predicts the observed results with a probability considered to be too low.

4.3   Transition probability

We consider, to simplify the discussion, a repeated experiment with a single possible outcome, which may have been realized or not after a given duration $t\tau$. Notice that this duration doesn't correspond to the instant at which some physical event occurs, but is a sufficiently long duration after which the experimenter can consciously remember of having (just) observed the expected outcome or not.

The possible mental states corresponding to the beginning of the experiment are written $\mathfrak{M}_i$ and the initial state of the quantum fields is also an element of the Hilbert subspace $\mathcal{H}_i$ given by:

$$\mathcal{H}_i = \overset{\perp}{\bigoplus_{\mathfrak{M}_i}} \mathcal{H}_{\mathfrak{M}_i}$$

The possible mental states corresponding to the measurement of the given outcome resp. of its absence are written $\mathfrak{M}_f^+$ resp. $\mathfrak{M}_f^-$. If the experiment works correctly, the final state of the quantum fields is, after measurement, an element of either of the Hilbert subspaces $\mathcal{H}_f^+$ or $\mathcal{H}_f^-$ given by:

$$\mathcal{H}_f^\pm = \overset{\perp}{\bigoplus_{\mathfrak{M}_f^\pm}} \mathcal{H}_{\mathfrak{M}_f^\pm}$$

If the experiment fails for some reason (e.g. if some measuring device is getting damaged during the experiment), the final state of the quantum fields is orthogonal to $\mathcal{H}_f^+ \oplus \mathcal{H}_f^-$.

The absolute probability of measuring the given outcome resp. its absence is given by:

$$\mathcal{P}\left(\mathcal{H}_i \to \mathcal{H}_f^\pm\right) = \sum_{\mathfrak{M}_{t-1}} \cdots \sum_{\mathfrak{M}_1} \mathrm{Tr}_{\mathcal{H}_i} \widehat{U}_\tau^\dagger \widehat{\Pi}_{\mathfrak{M}_1} \cdots \widehat{U}_\tau^\dagger \widehat{\Pi}_{f\pm} \widehat{U}_\tau \cdots \widehat{\Pi}_{\mathfrak{M}_1} \widehat{U}_\tau / \dim \mathcal{H}_i$$

where $\widehat{\Pi}_{f\pm} = \sum_{\mathfrak{M}_f^\pm} \widehat{\Pi}_{\mathfrak{M}_f^\pm}$. The conditional probability of measuring the given outcome if the experiment doesn't fail is then given by:

$$\mathcal{TP}\left(\mathcal{H}_i \to \mathcal{H}_f^+\right) = \frac{\mathcal{P}\left(\mathcal{H}_i \to \mathcal{H}_f^+\right)}{\mathcal{P}\left(\mathcal{H}_i \to \mathcal{H}_f^+\right) + \mathcal{P}\left(\mathcal{H}_i \to \mathcal{H}_f^-\right)}$$

and we call it "transition probability" from $\mathcal{H}_i$ to $\mathcal{H}_f^+$.

If the experiment is conceived in such a way that the studied system is isolated from the observer for the duration of the experiment until it interacts with some measurement apparatus, the experiment is considered to have failed if the observer has gained some information about the studied system before it interacts with this apparatus. An intermediate observation of the system, as it would leave a permanent trace in the memory of the observer, would lead with a vanishingly small probability to a final mental state in which the observer isn't conscious of having made this observation. The only intermediate mental states $\mathfrak{M}_1, \ldots, \mathfrak{M}_{t-1}$ to be considered in the above sums (*i.e.* which haven't a vanishingly small contribution to the transition probability) correspond also to projectors that don't affect the Hamiltonian evolution of the studied system. In that case, the absolute probability of measuring the given outcome resp. its absence can be approximated by:

$$\mathcal{P}\left(\mathcal{H}_i \to \mathcal{H}_f^{\pm}\right) \approx \mathrm{Tr}_{\mathcal{H}_i} \widehat{U}_{t\tau}^{\dagger} \widehat{\Pi}_{f^{\pm}} \widehat{U}_{t\tau} / \dim \mathcal{H}_i$$

and can be written as a sum resp. a mean on quantum states forming an orthonormal basis of $\mathcal{H}_f^{\pm}$ resp. $\mathcal{H}_i$:

$$\mathcal{P}\left(\mathcal{H}_i \to \mathcal{H}_f^{\pm}\right) \approx \sum_f \langle \mathcal{P}(i \to f) \rangle_i$$

$$\mathcal{P}(i \to f) := |U_{fi}(t\tau, 0)|^2$$

In this expression, the (absolute) transition probabilities $\mathcal{P}(i \to f)$ between two quantum states can be developed in series of the form:

$$\mathcal{P}(i \to f) = \sum_{n=0}^{\infty} \mathcal{P}^{(n)}(i \to f)$$

$$\mathcal{P}^{(n)}(i \to f) := \sum_{n_1+n_2=n} \overline{U_{fi}^{(n_1)}(t\tau, 0)} U_{fi}^{(n_2)}(t\tau, 0)$$

If i and f are plane wave states, these terms can be written using the scattering matrix as:

$$\mathcal{P}^{(n)}(i \to f) := \sum_{n_1+n_2=n} \overline{S_{fi}^{(n_1)}} S_{fi}^{(n_2)}$$

4.4   Leading order transition probability

The general form of the transition probability between plane wave modes of the field can be given without knowing much about the interaction term $\widehat{H}'$. We assume here that the initial and final states of the field are plane waves of the form:

$$|\Psi_i\rangle = \left|(N_{i\boldsymbol{q},\lambda}^{\phi})\right\rangle$$

$$|\Psi_f\rangle = \left|(N_{f\boldsymbol{q},\lambda}^{\phi})\right\rangle$$

The first interesting terms in the development of the transition probability are given in that case by:

$$\mathcal{P}^{(0)}(i \to f) := \overline{S_{fi}^{(0)}} S_{fi}^{(0)} = \delta_{f,i}$$

$$\mathcal{P}^{(1)}(i \to f) := \overline{S_{fi}^{(0)}} S_{fi}^{(1)} + \overline{S_{fi}^{(1)}} S_{fi}^{(0)} = 0$$

$$\mathcal{P}^{(2)}\left(i \to f\right) \quad := \quad \overline{S_{fi}^{(0)}}S_{fi}^{(2)} + \overline{S_{fi}^{(1)}}S_{fi}^{(1)} + \overline{S_{fi}^{(2)}}S_{fi}^{(0)}$$

$$= \quad (2\pi)^2 \frac{t - t_0}{h} \left|H_{f,i}'\right|^2 \delta_{t-t_0}^{(2)}\left(E_f - E_i\right)$$

$$-\delta_{f,i}(2\pi)^2 \frac{t - t_0}{h} \sum_{(N_{1\boldsymbol{q},\lambda}^{\phi})} \left|H_{1,i}'\right|^2 \delta_{t-t_0}^{(2)}\left(E_1 - E_i\right)$$

where the nascent delta function $\delta_{t-t_0}^{(2)}\left(E\right)$ is defined as in appendix A.3.

4.5 Higher order transition probability

To the order $n \geq 2$, the transition probability between plane wave states $\left|(N_{i\boldsymbol{q},\lambda}^{\phi})\right\rangle$ and $\left|(N_{f\boldsymbol{q},\lambda}^{\phi})\right\rangle$ is given by:

$$\mathcal{P}^{(n)}\left(i \to f\right) \quad := \quad \sum_{n_1+n_2=n} \overline{S_{fi}^{(n_1)}}S_{fi}^{(n_2)}$$

$$= \quad \delta_{f,i} \sum_{\substack{k=1 \\ (N_{k\boldsymbol{q},\lambda}^{\phi})}}^{n-1} \left(S_{i,\ldots,i}^{(n)} + \overline{S_{i,\ldots,i}^{(n)}}\right)$$

$$+ \sum_{\substack{n_1+n_2=n \\ n_1,n_2 \geq 1}} \overline{S_{fi}^{(n_1)}}S_{fi}^{(n_2)}$$

The first term vanishes for $f \neq i$. The development of the last term involves a "closed loop" of length n from i to i over f, *i.e.* a summation over $n - 2$ intermediate states $\left|(N_{k\boldsymbol{q},\lambda}^{\phi})\right\rangle$, where $k \in [\![-n_1, n_2]\!]$, $(N_{0\boldsymbol{q},\lambda}^{\phi}) = (N_{i\boldsymbol{q},\lambda}^{\phi})$ and $(N_{-n_1\boldsymbol{q},\lambda}^{\phi}) = (N_{n_2\boldsymbol{q},\lambda}^{\phi}) = (N_{f\boldsymbol{q},\lambda}^{\phi})$, and can be written as:

$$\sum_{\substack{n_1+n_2=n \\ n_1,n_2 \geq 1}} \sum_{\substack{k=-n_1 \\ (N_{k\boldsymbol{q},\lambda}^{\phi})}}^{n_2} \left(\prod_{k=-n_1}^{n_2-1} H_{k+1,k}'\right) \overline{S_{t-t_0}^{(n_1)}\left(E_{-n_1}, \ldots, E_0\right)} S_{t-t_0}^{(n_2)}\left(E_{n_2}, \ldots, E_0\right)$$

To the third order, for instance, the transition probability for $f \neq i$ reads:

$$\mathcal{P}^{(3)}\left(i \to f\right) = (2\pi)^3 \delta_{t-t_0}^{(1)}\left(E_f - E_i\right)$$

$$\sum_{(N_{1\boldsymbol{q},\lambda}^{\phi})} \left[\Im\left(H_{i,f}' H_{f,1}' H_{1,i}'\right) \delta_{t-t_0}^{(1)}\left(E_f - E_1\right) \delta_{t-t_0}^{(1)}\left(E_1 - E_i\right)\right.$$

$$\left.+\Re\left(H_{i,f}' H_{f,1}' H_{1,i}'\right) \frac{\delta_{t-t_0}^{(1)}\left(E_f - E_i\right) - \cos\left(\pi \frac{t-t_0}{h}(E_f - E_1)\right) \delta_{t-t_0}^{(1)}\left(E_1 - E_i\right)}{\pi(E_f - E_1)}\right]$$

where the nascent delta function $\delta_{t-t_0}^{(1)}\left(E\right)$ is defined as in appendix A.3.

4.6 Ideal experimental setup

We consider a scattering experiment designed to produce n_f particles of type ϕ_j, of wave vector $\boldsymbol{q}_j$ and of spin state λ_j. To detect them all, a set of n_f particle detectors D_j is being used and we consider a single alternative: either all the detectors are activated or at least one of them isn't. The momentum of the detected particles is measured with an uncertainty given by the domain δP_j of the momentum space in which particle j could be found without changing the measurement result. The corresponding subset δQ_j of values of $\boldsymbol{q}_j$ is given in the lattice reference frame by:

$$\delta Q_j = \left(\frac{[\![-N, N]\!]}{1+2N}\right)^3 \cap \frac{a}{h}\delta P_j$$

We assume that the corresponding subspace $\delta \mathcal{F}$ of $\mathcal{H}$ is given by:

$$\delta \mathcal{F} = \overset{\perp}{\underset{(\delta \boldsymbol{q}_j)}{\bigoplus}} \mathbb{C}\widehat{\Psi}^\dagger_{f+(\delta \boldsymbol{q}_j)}\,|O\rangle$$

$$\widehat{\Psi}^\dagger_{f+(\delta \boldsymbol{q}_j)} := \prod_j \widehat{a^{\phi_j}}^\dagger_{\boldsymbol{q}_j+\delta \boldsymbol{q}_j, \lambda_j}$$

where the summation goes over all the $\delta \boldsymbol{q}_j$ verifying $\boldsymbol{q}_j + \delta \boldsymbol{q}_j \in \delta Q_j$ and where $|O\rangle$ describes the experimental setup, including measuring devices and the observer.

The probability that all the detectors are activated is then given by:

$$\mathcal{P}\left(i \to \delta \mathcal{F}\right) := \sum_{(\delta \boldsymbol{q}_j)} \mathcal{P}\left(i \to f + (\delta \boldsymbol{q}_j)\right)$$

If the transition probability $\mathcal{P}\left(i \to f + (\delta \boldsymbol{q}_j)\right)$, as a function of $(\delta \boldsymbol{q}_j)$, admits a continuation on $\mathbb{R}^{3n_f}$, an approximation of this sum can be obtained by taking the corresponding integral:

$$\mathcal{P}\left(i \to \delta \mathcal{F}\right) \approx \int_{\prod_j \delta P_j} \mathcal{P}\left(i \to f + (\delta \boldsymbol{q}_j)\right) \left((1+2N)\frac{a}{h}\right)^{3n_f} d^3\boldsymbol{p}_1 \cdots d^3\boldsymbol{p}_{n_f}$$

where $\delta \boldsymbol{q}_j := \frac{a}{h}\boldsymbol{p}_j - \boldsymbol{q}_j$ in the lattice reference frame.

In particular, if $\left|H'_{f+(\delta \boldsymbol{q}_j),i}\right|^2$ admits such a continuation and if i and f could be approximated by plane wave states, the leading order transition probability could be approximated, for $i \notin \delta \mathcal{F}$, by:

$$\mathcal{P}^{(2)}\left(i \to \delta \mathcal{F}\right) \approx \int_{\prod_j \delta P_j} (2\pi)^2 \frac{t - t_0}{h} \left|H'_{f+(\delta \boldsymbol{q}_j),i}\right|^2 \delta^{(2)}_{t-t_0}\left(E_{f+(\delta \boldsymbol{q}_j)} - E_i\right)$$

$$\left((1+2N)\frac{a}{h}\right)^{3n_f} d^3\boldsymbol{p}_1 \cdots d^3\boldsymbol{p}_{n_f}$$

4.7   Quantum measurement

Let us consider a simple thought experiment in order to illustrate how measurement processes take place: An excited atom decays by emitting a photon, which is detected by a photomultiplier read by an observer. In the initial state, at time t_0, the atom has just been switched to its excited state and hasn't decayed yet, the measurement apparatus is indicating that it hasn't yet detected any photon and the observer is waiting for the detector to get activated. We symbolize this situation by:

The decay of the excited atom into a stable atom and a photon (via the QED processes described in chapter 8) first brings it into a quantum superposition of states, where both states coexist as a linear combination within the quantum state $|\psi\rangle$ of the universe. In both cases, the measurement apparatus is still inactive so far and the observer waiting. We symbolize this situation by:

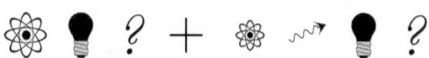

Supposing, to simplify, that the photomultiplier has a detection efficiency of 100%, it gets activated with certainty by the incoming photon after a certain delay Δt, coming thus itself into a quantum superposition of states. At the same time, the excited atom populates again the decayed atom state, as in the preceding step. In all three cases, the observer is still waiting that far. We symbolize this situation by:

Finally, the observer is becoming aware of the fact that the detector has been activated and comes herself into a quantum superposition of states, so that four qualitatively different states coexist as a linear combination within the quantum state $|\psi\rangle$ of the universe. We symbolize this situation by:

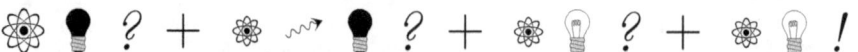

This (purely material) situation lasts until the next process of collapse and mental state selection takes place. The experienced consciousness state of the observer is then, randomly, either one of "I am still waiting for the detector to get activated" or "I have seen the detector becoming activated". In the first case, the quantum state of the universe becomes again:

i.e. the detector has still been activated, but the observer didn't yet notice it. In the second case, the state of the universe becomes:

and the atom has definitely decayed. The probability δP of this event to occur is very small, but the process of collapse and mental state selection take place very often, with a period τ, so that the decay of the atom should eventually be observed. The leading order approximation of this elementary probability takes the form:

$$\delta P \approx (2\pi)^2 \frac{\tau}{\text{h}} \sum_f \left| H'_{f,i} \right|^2 \delta^{(2)}_{t-t_0} \left(E_f - E_i \right)$$

where the Coulomb interaction term is supposed to have been shifted into the zeroth order Hamiltonian operator $\widehat{H}_0$, where the summation runs over an orthonormal basis of eigenstates of this operator and where the delta function can be approximated by the (time independent) density of decay states (with regard to energy) around the "allowed" decay states conserving zeroth order energy. The duration until the decay is being observed follows a Poisson law and its mean value is given by the general formula:

$$\langle t - t_0 \rangle \approx \Delta t + \tau \frac{1 - \delta P}{\delta P}$$

It can be approximated in this case, since $\delta P \ll 1$, by:

$$\langle t - t_0 \rangle \approx \Delta t + \text{h} \left[(2\pi)^2 \sum_f \left| H'_{f,i} \right|^2 \delta^{(2)}_{t-t_0} \left(E_f - E_i \right) \right]^{-1}$$

Note that this result doesn't depend on the period τ of the collapse and mental state selection process.

Part III

Interpretation

Chapter 5

Metaphysics

As I have said so many times,
God doesn't play dice with the world.

Albert Einstein,
in *Einstein and the Poet* [12]

5.1 Spinoza's philosophy

Since the interpretation of Quantum Field Theory I am about to give has been inspired by Spinoza's classical work *The Ethics* [18], I shall make here a short presentation of its basic ideas. According to the causalist world view of classical mechanics, each individual existent thing – an object, a thought – has necessarily a cause which explains its existence at a given moment. These things are considered to be alterations, or "modes", of some fundamental "substance" constitutive of Nature as a whole. Since this substance has some of the fundamental properties attributed to God by Judaic theology – self-caused, free, eternal, infinite (*i.e.* containing everything) –, it has been identified by Spinoza to God itself, confounding thus the concept of 'God' with what philosophers traditionally call 'Nature'. The human intellect conceives the substance, as well as every individual existent thing, under the two aspects, or "attributes", of an extended (material) and of a thinking (mental) thing. This categorization, however, is nothing but a property of the human intellect and not an intrinsic property of the things themselves. Considered under its material aspect, a human being, for instance, consists in a body extending in the substance, *i.e.* in God, whereas it consists in a mind thinking in God when considered under its mental aspect. Nevertheless, both are one and the same thing, so that the laws of Physics – considered to be part of the nature of God – could determine the laws of Psychology. The knowledge of God, which also encompasses the knowledge of the world in general and of Man in particular, is therefore considered to be the mind's highest good.

5.2   Quantum metaphysics

Interestingly, Spinoza's metaphysical concepts can be identified quite straightfor-wardly with the fundamental notions of Quantum Field Theory, thus providing them with a naturalistic basis. On the other hand, Quantum Field Theory, generally con-sidered to be counter-intuitive, paradoxical and hardly understandable, becomes grounded in a very classical philosophical tradition and should thus become accessi-ble to a broader range of Science philosophers.

The states (modes) of God (the substance) are evidently identified, under their material aspect, with the quantum states $|\Psi\rangle$ of the universe (the elements of the Hilbert space $\mathcal{H}$), and, under their mental aspect, with the mental states $\mathfrak{M}$ (the elements of $\mathcal{M}$). The relation between the material and the mental aspects is given by the decomposition $\mathcal{H} = \bigoplus_{\mathfrak{M}}^{\perp} \mathcal{H}_{\mathfrak{M}}$ of the Hilbert space, or equivalently by the orthogonal projection operators $\widehat{\Pi}_{\mathfrak{M}}$, relating each mental state $\mathfrak{M}$ with the set of all corresponding quantum states $\mathcal{H}_{\mathfrak{M}}$. Furthermore, the nature of God encompasses the laws of Physics, given by the Hamilton operator $\widehat{H}$, or more precisely by the elementary evolution operator $\widehat{U}_{\tau} := \exp\left(-\mathrm{i}2\pi\widehat{H}\tau/\mathrm{h}\right)$. God can finally be defined as a mathematical structure $\mathfrak{g}$ given by:

$$\mathfrak{g} := \left(\mathcal{H} \times \mathcal{M}, (\widehat{\Pi}_{\mathfrak{M}}), \widehat{U}_{\tau}\right)$$

The states of God, taking the general form ($|\Psi\rangle, \mathfrak{M}$), are said to be 'real' if $|\Psi\rangle \in \mathcal{H}_{\mathfrak{M}} \setminus \{0\}$ and 'virtual' otherwise. By extension, we will say that a quantum state $|\Psi\rangle \neq 0$ is 'real' if it belongs to one of the subspaces $\mathcal{H}_{\mathfrak{M}}$. An elementary evolution step of the state of God proceeds from a real state ($|\Psi\rangle_0, \mathfrak{M}_0$), first evolving to a generally virtual state $\left(\widehat{U}_{\tau} |\Psi_0\rangle, \mathfrak{M}_0\right)$ and eventually collapsing to one of the real states $\left(\widehat{\Pi}_{\mathfrak{M}_1}\widehat{U}_{\tau} |\Psi_0\rangle, \mathfrak{M}_1\right)$ with a probability $\langle\Psi_0| \widehat{U}_{\tau}^{\dagger}\widehat{\Pi}_{\mathfrak{M}_1}\widehat{U}_{\tau} |\Psi_0\rangle / \langle\Psi_0|\Psi_0\rangle$.

Chapter 6

Interpretation

6.1 The role of consciousness

As it results from the preceding description of the processes taking place in the evolution of the state of God, not only the purely material processes described by the Hamiltonian evolution operator $\widehat{U}_\tau$, but also the mental processes described by the collapse operators ($\widehat{\Pi}_{\mathfrak{M}}$) play a central role in the evolution of the quantum state $|\Psi\rangle$ of the universe. In the following, we will call 'consciousness state' any mental state $\mathfrak{M}$ and 'consciousness' the phenomenon of experiencing it. This phenomenon must be very carefully distinguished from the purely material processes of conscious thinking happening at the neural level within brains, although both are closely related to each other.

Quantum phenomena, like the superposition of an atom in an excited and a decayed state, or the superposition of a photodetector in an activated and an unactivated state – as in the quantum measurement example given in section 4.7 –, have always proved to be very puzzling to us, because they show that quantum processes don't fit within our mental categories, in which a photodetector must be either activated or not, for instance. This is not a result of a limitation of our intelligence that we could overcome by learning to think in a new way corresponding more adequately to the physical reality. No, at a very fundamental level, there isn't and there will never be any subjective experience $\mathfrak{m}$ corresponding to the superposition of a brain having "seen" a photodetector both activated and not, although this superposition does exist at the material level. This inadequacy between our mental categories and the material reality is a matter of fact having profound consequences for the process of consciousness. The contents of a consciousness state $\mathfrak{M}$ cannot simply be a representation of the material reality (even a partial one), because material reality explores possibilities going far beyond the realm of what we can grasp with our mental categories. An arbitrary consciousness state $\mathfrak{M}$ can only "match" more or less good the current quantum state $|\Psi\rangle$ of the universe, the number $\langle\Psi|\,\widehat{\Pi}_{\mathfrak{M}}\,|\Psi\rangle\,/\,\langle\Psi|\Psi\rangle$, lying between 0 and 1, measuring how good the fit is. If it equals 1, the fit is perfect (although $\mathfrak{M}$ remains a partial representation of the quantum state $|\Psi\rangle$) and $\mathfrak{M}$ is being experienced with certainty. If it equals 0, there is no fit and $\mathfrak{M}$ cannot be experienced. If it lies inbetween, any other consciousness state $\mathfrak{M}$ could be experienced too, the numbers $\langle \Psi | \ \widehat{\Pi}_{\mathfrak{M}} \ | \Psi \rangle \, / \, \langle \Psi | \Psi \rangle$ defining the probability law according to which the actually experienced consciousness state will be selected. So the quantum state $| \Psi \rangle$ of the universe determines the contents of the experienced consciousness state $\mathfrak{M}$, according to a probability law reflecting how good the mental categories in $\mathfrak{M}$ fit the material reality $| \Psi \rangle$. On the other hand, and this is probably even more astonishing, the experience of the consciousness state $\mathfrak{M}$ will reduce the quantum state $| \Psi \rangle$ of the universe to its component $\widehat{\Pi}_{\mathfrak{M}} \ | \Psi \rangle$ corresponding to the mental categories in $\mathfrak{M}$. So we can say that consciousness actively shapes the material reality according to its own mental categories – or more poetically, that you are putting human order into the world with every glance you take at it!

In the quantum measurement example given in section 4.7, for instance, the quantum superposition of two states of the brain of the observer, having either observed the photodetector activated or not, resolves to one of the components corresponding to the mental categories 'I have seen the photodetector activated' and 'I haven't seen it activated yet'. Because the quantum state of the rest of the universe (here specifically the photodetector and the decaying atom) is correlated to the quantum state of the brain via sensory perception, this reduction of the quantum state of the universe to a given mental category will have consequences for the rest of the universe, too. For instance, if the observer makes the conscious experience of seeing the photodetector activated, the state of the photodetector will also reduce to the activated state only, because the unactivated state is only correlated to the component of the state of the brain corresponding to the mental category 'I haven't seen the photodetector activated yet', which is being dropped. So the mental categories, which only concern the quantum state of the brain originally, get transposed to external objects via the correlation induced by sensory perception between them and the brain in the quantum state $| \Psi \rangle$ of the universe. Similarly, the state of the decaying atom will also reduce to the decayed state because the non-decayed state is only correlated to the component of the state of the brain corresponding to the mental category 'I haven't seen the photodetector activated yet', which is being dropped.

If the observer had made instead another kind of measurement on the decaying atom, e.g. measuring its position (which we suppose here to be uncorrelated with its decay), the quantum state of the atom would have been reduced accordingly, so that it would only be present in the region of space where it has been observed, whereas the decayed and non-decayed states would still remain in a quantum superposition, since they are not correlated with the consciousness state of the observer. So the way our mental categories get transposed to external objects strongly depends on the way we are observing them, *i.e.* on the way we are letting them get correlated to the quantum state of our brains.

6.2 Epistemological considerations

From a historical perspective, it should become quite clear today why the founders of Quantum Physics have had such difficulties to agree on an interpretation of this new branch of Physics. Starting with a few subatomic experiments, like the measurement of the emission spectrum of hydrogen atoms, they ended up with a theory bringing along a twofold scientific revolution and profoundly revising our world view.

The first revolution, which is nowadays widely accepted, concerns the fact that material reality cannot be described within our usual mental categories. The most classical example is the so-called wave-particle duality, which implies that elementary particles, and as a consequence also atoms and molecules for instance, can occupy several positions in space at a time and that their motion follow wave equations and interference patterns typical for wave phenomena. And ultimately, not only invisible particles, but also configurations of the whole universe can combine with each other via wave amplitudes and interfere in their evolution in a similar way as waves would do. This has been a big paradigm change compared to the ideal of Classical Physics, where intelligibility, *i.e.* the adequation to our mental categories, was considered an essential characteristic of any scientific theory. The position of Louis de Broglie, for instance, is typical for the efforts to resist this paradigm change. After having proposed himself the relation $\lambda = h/p$ between the momentum p of a particle and the wavelength λ of the corresponding wave phenomena, he developed the Pilot-Wave Theory, an alternative interpretation of Quantum Mechanics (that we know today to be false) according to which both the particle and the corresponding wave have their own existence and can be described as in Classical Physics, *i.e.* according to our mental categories – the particle having a definite trajectory and being "guided" by the accompanying wave.

The second scientific revolution, which is far from being over yet, concerns the fact that consciousness actively modifies the quantum state of the universe, according to its own mental categories and in its own way, which cannot be reduced to other, purely material phenomena: Technically, the collapse of a quantum state from $|\Psi\rangle$ to $\widehat{\Pi}_{\mathfrak{M}} |\Psi\rangle$ obviously cannot be reduced to a Hamiltonian evolution of the form $\widehat{U}_\tau |\Psi\rangle$. This is of course a radical paradigm change compared to the Cartesianism of the Copenhagen interpretation, according to which the consciousness of the "observer" passively takes notice of some aspects of the material world, e.g. the (supposedly well-defined) state of a measurement apparatus. A typical opponent to this paradigm change is Albert Einstein, who saw with very critical eyes the "spooky action at a distance" implied by the collapse of the quantum state of the universe, *i.e.* the instantaneous modification of the quantum state of a distant object happening when the consciousness state of a previously correlated brain is being selected. The famous Einstein-Podolsky-Rosen thought experiment, which has been conceived to illustrate these non-local features of collapse (*i.e.* its incompatibility to one of the central principles of Special Relativity), was thought to invalidate definitely the hypothesis of collapse, because it was conceived under the assumptions that all physical phenomena should obey the same laws, described in Quantum Theory by the Hamiltonian evolution operator $\widehat{U}_\tau$, and that collapse should ultimately be described in that way instead of using the ad-hoc assumption of the intervention of a projection operator $\widehat{\Pi}_{\mathfrak{M}}$. However, as this experiment has been realized for instance by the team of Alain Aspect [2] in experimental conditions becoming more and more sophisticated, the non-locality of quantum measurement has always been demonstrated very clearly, so that it makes no doubt today that collapse does obey other physical laws than Hamiltonian evolution alone.

The first milestone of this second scientific revolution has been set by John von Neumann with the idea that "mind causes collapse" [16]. This idea addresses a leak in the Copenhagen interpretation, where we distinguish between a "macroscopic world", supposed to be ruled by the laws of Classical Physics, and a "microscopic world", ruled by the laws of Quantum Physics. The interface between both worlds is build by measurement apparatuses, which are supposed to cause the collapse of the quantum state of the microscopic world when they interact with it. This interpretation relies on the assumption that no quantum phenomenon can be observed without the help of a measurement apparatus, which is obviously false. For instance, you can observe with your naked eyes the diffraction patterns of light passing in the dark through the fine structures of woven fabric, and that is a genuine quantum phenomenon. One could maybe "save" the Copenhagen interpretation by considering that the eye constitute the measurement apparatus in that case, but where do an eye actually begin: With the cornea, the pupil, the retina, the retinal ganglion cells? Or even inside the brain, after the neural processing of visual perceptions? Defining the frontier between the microscopic and the macroscopic world seems to be a rather arbitrary operation and it is therefore not really intellectually satisfying. The only thing we are sure of is that, ultimately, our consciousness "resolves" quantum superpositions according to our mental categories. This idea that, ultimately, consciousness causes collapse was once known as the 'standard interpretation' of Quantum Mechanics. It has been almost forgotten since, perhaps because it had been originally formulated all to vaguely to be taken seriously. In this book, I am formulating it again using a very precise and well-defined formalism, so that one can unambiguously derive its implications on a very solid basis. I hope that this contribution will help reconsidering the profound implications of this second scientific revolution and widening its acceptance in the scientific community.

6.3 The Spinozist approach to Quantum Physics

The Spinozist aspects of our interpretation of Quantum Field Theory concern this second scientific revolution, *i.e.* the role of consciousness in physical processes. Historically, Spinoza's philosophy developed on top of Cartesianism, which considers the material world to be a mechanical, deterministic one and consciousness to be a passive, external observer of the happenings in the material world; although this material world is supposed to obey the very strict laws of Classical Mechanics, the mental world is supposed to be absolutely free, independent of the material one and obeying no specific laws. Spinoza conserved this mechanical view of the material world, but tried to ground the mental world upon the material one, considering that consciousness cannot exist independently of a material body, that it reflects the state of its material substrate and therefore obeys the same laws, which can be transposed, in principle, to the mental world. Thus, consciousness becomes again part of Nature; it isn't considered any more to exist in an ideal realm exterior to the contingencies of the material world.

Our interpretation of Quantum Field Theory relates to the Copenhagen interpretation in a similar way as Spinozism relates to Cartesianism. In the Copenhagen interpretation, the microscopic world only – and only the limited system under consideration – is supposed to obey the laws of Quantum Physics, *i.e.* the Hamiltonian evolution and the collapse as the system interacts with a measurement apparatus. On the contrary, the macroscopic world, including the observer, is supposed to exist in an ideal realm where only the (Cartesian) laws of Classical Physics apply. In a genuine Spinozist approach, our interpretation grounds this ideal realm upon the material realm of the quantum world: The arbitrary distinction between a microscopic and a macroscopic world vanishes, whereas collapse is supposed not to happen in an interaction with a measurement apparatus – a mere artifact – but with brains – the material substrate of a fundamental aspect of Nature, consciousness. Mind becomes thus again part of Nature, and comes along with its own properties and physical laws, completing the Hamiltonian evolution laws of purely material processes. Of course, the relation between mind and body is much more complex than in classical Spinozism, but I think that the basic approach of the problem of consciousness is essentially the same, so that we can say, in that sense, that we are developing here a Spinozist interpretation of Quantum Physics.

Chapter 7

Philosophical issues

7.1 The explanatory gap

As it has been observed by science philosophers in the last century, the gap between our understanding level of physical-material and of mental phenomena has been continuously growing as the scientific community successfully focused on the development of the Relativity and Quantum theories. It is therefore quite understandable on a science psychological level that some cognitive neuroscientists may have been hoping to be able to explain one day all mental phenomena in terms of biophysical processes. However, even if we could describe one day the correspondence between mental states and quantum states of the brain, the question of knowing "why" some specific aspects of the material world correspond to certain mental experiences, and why some other aspects do not, would still remain open. This question is known as the "explanatory gap" and isn't to be answered by a theory focusing uniquely on the material world. The theory developed in this book addresses this issue in a threefold way. First, it gives a well-defined status to mental states, considered to be an aspect of reality on their own that isn't merely derived from the material one. This is expressed in the theory by the form $\mathcal{H} \times \mathcal{M}$ of the set of the possible states of God. Second, it defines the form of the correspondence between material and mental states, which is given by the family of the supplementary subspaces $\mathcal{H}_\mathfrak{M}$ corresponding to each possible mental state $\mathfrak{M}$. This stresses the idea that individual subjective experiences are not necessarily only related to material aspects of a single, individual brain, but that the totality of all subjective experiences globally relates to the quantum state of the universe. Finally, the description of the random collapse process from a virtual into a real state of God gives a first explanation of what is happening at a material and at a mental level as a mental state is getting experienced.

7.2 Skepticism

According to philosophical skepticism, in the form of Descartes' *Cogito Ergo Sum* argument in his *Discourse on the Method* [6] for instance, the one and only aspect of the world which we know beyond any doubt to be real is our present subjective experience, the 'Cogito'. Nothing can guarantee us that the representations of the world it carries – like our past experiences, the feeling of the permanence of our existence, the image of our body, of the outer world, of our relations to others – have or have had any physical reality. In particular, it cannot be taken for granted that experimental evidence can be accumulated over the ages: Experimental science *must* rely on the mere belief that the mental representations of what we consider to be accumulated experimental evidence are related to physical processes that really did happen in the past. Indeed, in physical terms, stating that I am having some subjective experience $\mathfrak{m}_s$ only implies that the mental state $\mathfrak{M} = (N_\mathfrak{m})$ is such that $N_{\mathfrak{m}_s} \geq 1$ and that the quantum state $|\Psi\rangle$ of the universe belongs to $\mathcal{H}^+_{\mathfrak{m}_s}$. It doesn't necessarily imply that the past evolution of $|\Psi\rangle$ corresponds to the mental representation of past events in the subjective experience $\mathfrak{m}_s$. The physical theory presented in this book belongs therefore to the long tradition of philosophical skepticism insofar as it doubts the very possibility of experimental science.

7.3 Materialism

Materialism is the doctrine according to which the subjective experience of consciousness can be completely reduced to the corresponding physical-material processes happening within our brains and thus can be explained without involving any other level of reality than the purely material one. It is generally considered among philosophers as the daydream of a physicist absorbed by his study object and becoming blind for the reality of his own subjective experience. Nevertheless, it still has numerous supporters in today's scientific community. In the frame of the theory developed in this book, it could be formulated as the hypothesis that no subjective experience is possible, since this would be equivalent to denying the existence of the mental world, which is of another nature as the material one. Mathematically, this hypothesis can be expressed simply as $\mathcal{H}_{\mathfrak{M}_\Omega} = \mathcal{H}$, so that no subject is having any subjective experience in any quantum state. Equivalently, this could be expressed in terms of collapse operators by $\widehat{\Pi}_{\mathfrak{M}_\Omega} = \mathbb{1}$, so that there is no collapse of the quantum state of the universe. Its evolution reduces therefore to its Hamiltonian part,

$$|\Psi(t)\rangle = \exp\left(-\mathrm{i}2\pi\frac{t-t_0}{\mathrm{h}}\widehat{\mathrm{H}}\right)|\Psi(t_0)\rangle$$

and the stochastic process of mental state selection do not apply.

Materialism in this context is facing the problem that it cannot satisfactorily explain how it is supposed to "feel like" in quantum states where brains happen to be in a quantum superposition of states corresponding to different states of consciousness. This would be the case for instance in a state of the form:

$$\left(\sqrt{0.9}\,\widehat{\Psi^\alpha_\mathfrak{m}}^\dagger + \sqrt{0.1}\,\widehat{\Psi^{\alpha'}_{\mathfrak{m}'}}^\dagger\right)\mathcal{H}_{\mathfrak{M}_\Omega}$$

where the brain states corresponding to the mental states 1_m and $1_{m'}$ are both present in a quantum superposition with the statistical weights 90% and 10%, respectively. There are two well-known ways of trying to escape this issue. In the no collapse theory of Everett, each consciousness state in the quantum superposition of a brain is supposed to be equally real as the others and to be experienced on its own. More precisely, these consciousness states are supposed to be statistically "weighted" in some (mysterious) way (since there isn't any random process taking place) by the square norm of the corresponding component of the quantum state of the universe, so that we are supposed to be more likely to experience one of them if it corresponds to a component with a greater square norm.

The second way of escaping the difficulties of materialism is to deny that there are "noticeable" quantum superpositions of consciousness states of the brain. This is basically the aim of all spontaneous collapse theories, which have been reviewed exhaustively by Angelo Bassi and GianCarlo Ghirardi in their report *Dynamical Reduction Models* [3]. Generally, the quantum state of the universe is supposed to collapse in such a way that the center of mass of macroscopic objects is practically always localized in a small region of space, so that we cannot notice its quantum fluctuations with our naked senses. As a consequence, insofar as our consciousness state is being mostly driven by sensory experience only, the states of consciousness corresponding to the components of a quantum superposition of brains are most likely to differ very little from another, so that it shouldn't really mind if we don't know exactly which one is being experienced.

7.4 Solipsism

The solipsist is convinced that she is (and must be) the only person in the universe who has a subjective experience. Solipsism makes thus unproblematic the fact that we are experiencing the mental world in the form of a single subjective experience instead of experiencing the whole mental state directly. In the frame of the theory developed in this book, solipsism can be expressed as the hypothesis that the only possible mental states (apart from $\mathfrak{M}_\Omega$) are of the form 1_m, or in physical terms, that:

$$\mathcal{H} = \mathcal{H}_{\mathfrak{M}_\Omega} \overset{\perp}{\oplus} \overset{\perp}{\bigoplus_m} \mathcal{H}_{1_m}$$

Of course, this hypothesis is logically perfectly correct, but it is utmost difficult to make it compatible with the idea that mental states are being realized physically by the presence of corresponding quantum states of brains. Even if one supposes that the solipsist's brain has something special that makes it differ from other brains that aren't giving rise to a subjective experience, one faces the problem that a quantum state in which many "copies" of the solipsist's brain, corresponding to different subjective experiences, would be present couldn't be related in a satisfactory way to a single subjective experience: It is unclear, for instance, if quantum states in a subspace of the form $\widehat{\Psi_m^{\alpha'}}^\dagger \widehat{\Psi_m^{\alpha}}^\dagger \mathcal{H}_{\mathfrak{M}_\Omega}$ should be experienced as 1_m or $1_{m'}$.

7.5 Soul immortality theorem

REMINDER As stated in section 4.2, Quantum Field Theory defines, for an arbitrary initial quantum state $|\Psi_i\rangle \neq 0$ realized at the initial time $t_i = 0$, the probability laws $P_t(\mathfrak{M}_0, \ldots, \mathfrak{M}_t; |\Psi_i\rangle)$, where $t \in \mathbb{N}$, that a given sequence of mental states $\mathfrak{M}_0, \ldots, \mathfrak{M}_t$ is being experienced at times $0, \ldots, t\tau$. These probability laws read:

$$P_t(\mathfrak{M}_0, \ldots, \mathfrak{M}_t; |\Psi_i\rangle) = \langle \Psi_i | \; \widehat{\Pi}_{\mathfrak{M}_0} \widehat{U}_\tau^\dagger \widehat{\Pi}_{\mathfrak{M}_1} \cdots \widehat{U}_\tau^\dagger \widehat{\Pi}_{\mathfrak{M}_t} \widehat{U}_\tau \cdots \widehat{\Pi}_{\mathfrak{M}_1} \widehat{U}_\tau \widehat{\Pi}_{\mathfrak{M}_0} \; |\Psi_i\rangle \, / \, \langle \Psi_i | \Psi_i \rangle$$

DEFINITIONS An infinite sequence $(\mathfrak{M}_t)$, indexed on $t \in \mathbb{N}$, is said to be "dead" if it is constant, *i.e.* if $\mathfrak{M}_t = \mathfrak{M}_0$ for all $t \in \mathbb{N}$; to be "eventually dying" if it becomes constant after a certain point, *i.e.* if there exists a $t_1 \in \mathbb{N}^*$ such that $\mathfrak{M}_t = \mathfrak{M}_{t_1}$ for all $t \geq t_1$, although $\mathfrak{M}_{t_0} \neq \mathfrak{M}_{t_1}$ for some $t_0 < t_1$; and to be "immortal" otherwise.

It is said to be "certain", for a given initial quantum state $|\Psi_i\rangle \neq 0$, if we have $P_t(\mathfrak{M}_0, \ldots, \mathfrak{M}_t; |\Psi_i\rangle) = 1$ for all $t \in \mathbb{N}$; to be "infinitely improbable" if $\lim_{t\to\infty} P_t(\mathfrak{M}_0, \ldots, \mathfrak{M}_t; |\Psi_i\rangle) = 0$; and to be "contingent" otherwise.

A quantum state is said to be "certainly dead" if, taken as an initial state, one of the dead sequences is certain; to be "possibly dead" if a dead sequence is contingent; to be "mortal" if an eventually dying sequence is certain or at least contingent; and to be "immortal" otherwise.

LEMMA Eventually dying sequences are infinitely improbable.

PROOF An eventually dying sequence $(\mathfrak{M}_t)$ is characterized, per definition, by the existence of a $t_f \in \mathbb{N}^*$ such that:

$$t_f = \min\{t \in \mathbb{N} \mid \forall t' > t, \; \mathfrak{M}_{t'} = \mathfrak{M}_t\}$$

This sequence is, per definition, infinitely improbable if and only if, for any initial quantum state $|\Psi_i\rangle \neq 0$, we have:

$$P_t(\mathfrak{M}_0, \ldots, \mathfrak{M}_t; |\Psi_i\rangle) \xrightarrow{t\to\infty} 0$$

The analysis of the asymptotic behavior can be simplified by using, for $t \geq t_f$, following factorization:

$$P_t(\mathfrak{M}_0, \ldots, \mathfrak{M}_t; |\Psi_i\rangle) = P_{t_f}(\mathfrak{M}_0, \ldots, \mathfrak{M}_{t_f}; |\Psi_i\rangle)$$
$$P_{t-t_f}(\mathfrak{M}_{t_f}, \ldots, \mathfrak{M}_t; \widehat{\Pi}_{\mathfrak{M}_{t_f}} \widehat{U}_\tau \cdots \widehat{\Pi}_{\mathfrak{M}_1} \widehat{U}_\tau \widehat{\Pi}_{\mathfrak{M}_0} |\Psi_i\rangle)$$

The asymptotic behavior is governed by the second factor, which we will analyze under the generic form:

$$P_t(\mathfrak{M}, \ldots, \mathfrak{M}; |\Psi\rangle) = \langle \Psi | \; (\widehat{\Pi}_{\mathfrak{M}} \widehat{U}_\tau^\dagger)^t \widehat{\Pi}_{\mathfrak{M}} (\widehat{U}_\tau \widehat{\Pi}_{\mathfrak{M}})^t \; |\Psi\rangle \, / \, \langle \Psi | \Psi \rangle$$

We will show that it admits a limit of the form:

$$P_t(\mathfrak{M}, \ldots, \mathfrak{M}; |\Psi\rangle) \xrightarrow{t\to\infty} \langle \Psi | \; \widehat{\Pi}_{\mathcal{H}_{\mathfrak{M}}^\infty} \; |\Psi\rangle \, / \, \langle \Psi | \Psi \rangle$$

where $\mathcal{H}_{\mathfrak{M}}^\infty$ is a subspace of $\mathcal{H}_{\mathfrak{M}}$ that we will define subsequently. We have therefore:

$$P_t(\mathfrak{M}_0, \ldots, \mathfrak{M}_t; |\Psi_i\rangle) \xrightarrow{t\to\infty}$$
$$\langle \Psi_i | \; \widehat{\Pi}_{\mathfrak{M}_0} \widehat{U}_\tau^\dagger \widehat{\Pi}_{\mathfrak{M}_1} \cdots \widehat{U}_\tau^\dagger \widehat{\Pi}_{\mathcal{H}_{\mathfrak{M}_{t_f}}^\infty} \widehat{U}_\tau \cdots \widehat{\Pi}_{\mathfrak{M}_1} \widehat{U}_\tau \widehat{\Pi}_{\mathfrak{M}_0} \; |\Psi_i\rangle \, / \, \langle \Psi_i | \Psi_i \rangle$$

and proving that $\widehat{\Pi}_{\mathcal{H}_{\mathfrak{M}}^{\infty}} \widehat{U}_{\tau} \widehat{\Pi}_{\mathfrak{M}'} = 0$ for any $\mathfrak{M}' \neq \mathfrak{M}$ will yield the conclusion.

§ Let us first analyze the structure of a mental subspace $\mathcal{H}_{\mathfrak{M}}$. Let $\mathcal{H}_{\mathfrak{M}}^{n}$ be the subspace of the quantum states in $\mathcal{H}_{\mathfrak{M}}$ remaining in $\mathcal{H}_{\mathfrak{M}}$ after $1, \ldots, n$ applications of the elementary evolution operator $\widehat{U}_{\tau}$. This operator being unitary, it is invertible, so that we can define these subspaces by:

$$\mathcal{H}_{\mathfrak{M}}^{n} := \bigcap_{t=0}^{n} \widehat{U}_{\tau}^{-t} \mathcal{H}_{\mathfrak{M}}$$

where $n \in \mathbb{N}$. These subspaces are included in each other by construction, and we write $\mathcal{H}_{\mathfrak{M}}^{\infty}$ their intersection, given by:

$$\mathcal{H}_{\mathfrak{M}}^{\infty} := \bigcap_{t=0}^{\infty} \widehat{U}_{\tau}^{-t} \mathcal{H}_{\mathfrak{M}}$$

§ $\mathcal{H}_{\mathfrak{M}}$ being finite dimensional, there must exist a finite number of distinct nested subspaces $\mathcal{H}_{\mathfrak{M}}^{n}$. Indeed, $\widehat{U}_{\tau}^{-1}$ being injective, we have

$$\mathcal{H}_{\mathfrak{M}}^{n+1} = \mathcal{H}_{\mathfrak{M}} \cap \widehat{U}_{\tau}^{-1} \mathcal{H}_{\mathfrak{M}}^{n}$$

for any $n \in \mathbb{N}$; if there exists a $n \in \mathbb{N}$ such that $\mathcal{H}_{\mathfrak{M}}^{n} = \mathcal{H}_{\mathfrak{M}}^{n+1}$, we have therefore by recurrence $\mathcal{H}_{\mathfrak{M}}^{n} = \mathcal{H}_{\mathfrak{M}}^{n'} = \mathcal{H}_{\mathfrak{M}}^{\infty}$ for any $n' \geq n$. If there didn't exist such a $n \in \mathbb{N}$, the inclusions $\mathcal{H}_{\mathfrak{M}}^{n+1} \subset \mathcal{H}_{\mathfrak{M}}^{n}$ would all be strict, so that the dimension of $\mathcal{H}_{\mathfrak{M}}^{0} = \mathcal{H}_{\mathfrak{M}}$ would be infinite, which isn't the case. There must also exist a $n_{\mathfrak{M}} \in \mathbb{N}$ such that

$$n_{\mathfrak{M}} = \min\{n \in \mathbb{N} \mid \mathcal{H}_{\mathfrak{M}}^{n} = \mathcal{H}_{\mathfrak{M}}^{n+1}\}$$

and we have $\mathcal{H}_{\mathfrak{M}}^{n_{\mathfrak{M}}} = \mathcal{H}_{\mathfrak{M}}^{\infty}$.

§ In the special case $n_{\mathfrak{M}} = 0$, it is trivial to show that $P_t(\mathfrak{M}, \ldots, \mathfrak{M}; |\Psi\rangle)$ tends to $\langle\Psi| \widehat{\Pi}_{\mathcal{H}_{\mathfrak{M}}^{\infty}} |\Psi\rangle / \langle\Psi|\Psi\rangle$ for any initial quantum state $|\Psi\rangle \neq 0$. In this case, we have $\mathcal{H}_{\mathfrak{M}}^{\infty} = \mathcal{H}_{\mathfrak{M}}$ and we will see that $P_t(\mathfrak{M}, \ldots, \mathfrak{M}; |\Psi\rangle) = \langle\Psi| \widehat{\Pi}_{\mathcal{H}_{\mathfrak{M}}^{\infty}} |\Psi\rangle / \langle\Psi|\Psi\rangle$ for any $t \in \mathbb{N}$. The assertion holds for $t = 0$ per definition, and $\widehat{\Pi}_{\mathfrak{M}} |\Psi\rangle \in \mathcal{H}_{\mathfrak{M}}$ since $\widehat{\Pi}_{\mathfrak{M}}$ is a projection operator on $\mathcal{H}_{\mathfrak{M}}$. For any quantum state $|\Psi_{\mathfrak{M}}\rangle \in \mathcal{H}_{\mathfrak{M}}$, $\widehat{U}_{\tau} |\Psi_{\mathfrak{M}}\rangle$ has the same norm as $|\Psi_{\mathfrak{M}}\rangle$ because of the unitarity of $\widehat{U}_{\tau}$, and since $\mathcal{H}_{\mathfrak{M}} = \mathcal{H}_{\mathfrak{M}}^{\infty} \subset \widehat{U}_{\tau}^{-1} \mathcal{H}_{\mathfrak{M}}$, $\widehat{U}_{\tau} |\Psi_{\mathfrak{M}}\rangle$ belongs to $\mathcal{H}_{\mathfrak{M}}$, so that $\widehat{\Pi}_{\mathfrak{M}} \widehat{U}_{\tau} |\Psi_{\mathfrak{M}}\rangle$ is equal to $\widehat{U}_{\tau} |\Psi_{\mathfrak{M}}\rangle$ and has the same norm as $|\Psi_{\mathfrak{M}}\rangle$, too. It is easy then to prove by recurrence that $(\widehat{\Pi}_{\mathfrak{M}} \widehat{U}_{\tau})^{t} |\Psi_{\mathfrak{M}}\rangle$ is equal to $\widehat{U}_{\tau}^{t} |\Psi_{\mathfrak{M}}\rangle$ and has therefore the same norm as $|\Psi_{\mathfrak{M}}\rangle$ for any $t \in \mathbb{N}$, so that we have:

$$\begin{aligned} P_t(\mathfrak{M}, \ldots, \mathfrak{M}; |\Psi\rangle) &= \langle\Psi| \widehat{\Pi}_{\mathfrak{M}} (\widehat{U}_{\tau}^{\dagger} \widehat{\Pi}_{\mathfrak{M}})^{t} (\widehat{\Pi}_{\mathfrak{M}} \widehat{U}_{\tau})^{t} \widehat{\Pi}_{\mathfrak{M}} |\Psi\rangle / \langle\Psi|\Psi\rangle \\ &= \langle\Psi| \widehat{\Pi}_{\mathfrak{M}} |\Psi\rangle / \langle\Psi|\Psi\rangle = \langle\Psi| \widehat{\Pi}_{\mathcal{H}_{\mathfrak{M}}^{\infty}} |\Psi\rangle / \langle\Psi|\Psi\rangle \end{aligned}$$

for any quantum state $|\Psi\rangle \neq 0$ and any $t \in \mathbb{N}$.

§ We assume from now on $n_{\mathfrak{M}} \geq 1$, so that the supplementary subspace $\mathcal{H}_{\mathfrak{M}}^{\infty\perp}$ of $\mathcal{H}_{\mathfrak{M}}^{\infty}$ in $\mathcal{H}_{\mathfrak{M}}$, defined by:

$$\mathcal{H}_{\mathfrak{M}} = \mathcal{H}_{\mathfrak{M}}^{\infty} \overset{\perp}{\oplus} \mathcal{H}_{\mathfrak{M}}^{\infty\perp}$$

isn't reduced to the zero subspace. We will consider the endomorphism $\widehat{U}_{\mathfrak{M}}$ induced by $\widehat{U}_{\tau}$ on $\mathcal{H}_{\mathfrak{M}}$, given by:

$$\widehat{U}_{\mathfrak{M}} := \widehat{\Pi}_{\mathfrak{M}} \widehat{U}_{\tau} \widehat{\Pi}_{\mathfrak{M}}$$

With this operator, we can write for any $t \in \mathbb{N}^*$ and any quantum state $|\Psi\rangle \neq 0$:

$$\mathrm{P}_t(\mathfrak{M}, \ldots, \mathfrak{M}; |\Psi\rangle) = \langle\Psi| \, \widehat{U}_{\mathfrak{M}}^{\dagger}{}^t \widehat{U}_{\mathfrak{M}}^t \, |\Psi\rangle / \langle\Psi|\Psi\rangle$$

and we will see that the subspaces $\mathcal{H}_{\mathfrak{M}}^n$ can be characterized, for any $n \in \mathbb{N}^*$, by:

$$\mathcal{H}_{\mathfrak{M}}^n = \{ \, |\Psi\rangle \in \mathcal{H} \mid \langle\Psi| \, \widehat{U}_{\mathfrak{M}}^{\dagger}{}^n \widehat{U}_{\mathfrak{M}}^n \, |\Psi\rangle = \langle\Psi|\Psi\rangle \}$$

§ Let us prove it by recurrence. $\widehat{\Pi}_{\mathfrak{M}}$ being the orthogonal projection operator on $\mathcal{H}_{\mathfrak{M}}$, for any quantum state $|\Psi\rangle \in \mathcal{H}$, $\widehat{\Pi}_{\mathfrak{M}} |\Psi\rangle$ and $|\Psi\rangle$ have the same norm if and only if $|\Psi\rangle \in \mathcal{H}_{\mathfrak{M}}$, otherwise $\langle\Psi| \, \widehat{\Pi}_{\mathfrak{M}} \, |\Psi\rangle < \langle\Psi|\Psi\rangle$. $\widehat{U}_{\tau}$ being unitary, it preserves the norm of $\widehat{\Pi}_{\mathfrak{M}} |\Psi\rangle$. Consequently, $\widehat{U}_{\mathfrak{M}} |\Psi\rangle$ and $|\Psi\rangle$ have the same norm if and only if $|\Psi\rangle \in \mathcal{H}_{\mathfrak{M}}$ and $\widehat{U}_{\tau}\widehat{\Pi}_{\mathfrak{M}} |\Psi\rangle \in \mathcal{H}_{\mathfrak{M}}$, otherwise $\langle\Psi| \, \widehat{U}_{\mathfrak{M}}^{\dagger}\widehat{U}_{\mathfrak{M}} \, |\Psi\rangle < \langle\Psi|\Psi\rangle$. For any $|\Psi\rangle \in \mathcal{H}_{\mathfrak{M}}$, $\widehat{\Pi}_{\mathfrak{M}} |\Psi\rangle = |\Psi\rangle$, so that this condition is equivalent to $|\Psi\rangle \in \mathcal{H}_{\mathfrak{M}} \cap \widehat{U}_{\tau}^{-1}\mathcal{H}_{\mathfrak{M}} = \mathcal{H}_{\mathfrak{M}}^1$. We have therefore:

$$\mathcal{H}_{\mathfrak{M}}^1 = \{ \, |\Psi\rangle \in \mathcal{H} \mid \langle\Psi| \, \widehat{U}_{\mathfrak{M}}^{\dagger}\widehat{U}_{\mathfrak{M}} \, |\Psi\rangle = \langle\Psi|\Psi\rangle \}$$

which proves the assertion for $n = 1$. Let us assume that, for a given $n \in \mathbb{N}^*$, $\widehat{U}_{\mathfrak{M}}^n |\Psi\rangle$ and $|\Psi\rangle$ have the same norm if and only if $|\Psi\rangle \in \mathcal{H}_{\mathfrak{M}}^n$, whereas $\langle\Psi| \, \widehat{U}_{\mathfrak{M}}^{\dagger}{}^n \widehat{U}_{\mathfrak{M}}^n \, |\Psi\rangle < \langle\Psi|\Psi\rangle$ otherwise. It can then be proved like above that $\widehat{U}_{\mathfrak{M}}^{n+1} |\Psi\rangle$ and $|\Psi\rangle$ have the same norm if and only if $|\Psi\rangle \in \mathcal{H}_{\mathfrak{M}}^n$, $\widehat{U}_{\mathfrak{M}}^n |\Psi\rangle \in \mathcal{H}_{\mathfrak{M}}$ and $\widehat{U}_{\tau}\widehat{U}_{\mathfrak{M}}^n |\Psi\rangle \in \mathcal{H}_{\mathfrak{M}}$, whereas $\langle\Psi| \, \widehat{U}_{\mathfrak{M}}^{\dagger}{}^{n+1} \widehat{U}_{\mathfrak{M}}^{n+1} \, |\Psi\rangle < \langle\Psi|\Psi\rangle$ otherwise. For any $|\Psi\rangle \in \mathcal{H}_{\mathfrak{M}}^n$, $\widehat{U}_{\mathfrak{M}}^n |\Psi\rangle = \widehat{U}_{\tau}^n |\Psi\rangle$, so that this condition is equivalent to $|\Psi\rangle \in \mathcal{H}_{\mathfrak{M}}^n \cap \widehat{U}_{\tau}^{-n}\mathcal{H}_{\mathfrak{M}} \cap \widehat{U}_{\tau}^{-(n+1)}\mathcal{H}_{\mathfrak{M}} = \mathcal{H}_{\mathfrak{M}}^{n+1}$, which proves the assertion for any $n \in \mathbb{N}^*$ by recurrence.

§ As a consequence, since $\mathcal{H}_{\mathfrak{M}}^{n_{\mathfrak{M}}} = \mathcal{H}_{\mathfrak{M}}^{\infty}$ and $\mathcal{H}_{\mathfrak{M}}^{\infty\perp}$ are disjoint by construction, we have in particular:

$$\forall \, |\Psi\rangle \in \mathcal{H}_{\mathfrak{M}}^{\infty\perp} \setminus \{0\}, \; \mathrm{P}_{n_{\mathfrak{M}}}(\mathfrak{M}, \ldots, \mathfrak{M}; |\Psi\rangle) < 1$$

The function $\mathrm{P}_{n_{\mathfrak{M}}}(\mathfrak{M}, \ldots, \mathfrak{M}; |\Psi\rangle)$ is continuous on $\mathcal{H} \setminus \{0\}$ and, being constant on the rays $\mathbb{C}^* \, |\Psi\rangle$, its maximum on $\mathcal{H}_{\mathfrak{M}}^{\infty\perp} \setminus \{0\}$ can be evaluated on the unit sphere of this subspace. Because of the finite dimensionality of $\mathcal{H}_{\mathfrak{M}}^{\infty\perp}$, the unit sphere is compact, so that this maximum exists and is being reached. There exists therefore a $p_{\mathfrak{M}} \in [0, 1[$ such that:

$$p_{\mathfrak{M}}^{n_{\mathfrak{M}}} = \max\{\mathrm{P}_{n_{\mathfrak{M}}}(\mathfrak{M}, \ldots, \mathfrak{M}; |\Psi\rangle) \mid |\Psi\rangle \in \mathcal{H}_{\mathfrak{M}}^{\infty\perp} \setminus \{0\}\}$$

This will allow us to set an upper bound to $\mathrm{P}_t(\mathfrak{M}, \ldots, \mathfrak{M}; |\Psi\rangle)$ for any $t \geq n_{\mathfrak{M}}$. Per factorization, for any quantum state $|\Psi\rangle \in \mathcal{H} \setminus \{0\}$ and any $k \in \mathbb{N}^*$, we have:

$$\mathrm{P}_{k n_{\mathfrak{M}}}(\mathfrak{M}, \ldots, \mathfrak{M}; |\Psi\rangle) = \mathrm{P}_{n_{\mathfrak{M}}}(\mathfrak{M}, \ldots, \mathfrak{M}; |\Psi\rangle)^k$$

and, $\mathrm{P}_t(\mathfrak{M}, \ldots, \mathfrak{M}; |\Psi\rangle)$ being a decreasing function of t, we have more generally for any $t \geq n_{\mathfrak{M}}$:

$$\mathrm{P}_t(\mathfrak{M}, \ldots, \mathfrak{M}; |\Psi\rangle) \leq \mathrm{P}_{n_{\mathfrak{M}}}(\mathfrak{M}, \ldots, \mathfrak{M}; |\Psi\rangle)^{\lfloor t/n_{\mathfrak{M}} \rfloor}$$

where $\lfloor \cdot \rfloor$ denotes the floor function. In particular, for any quantum state $|\Psi\rangle \in \mathcal{H}_{\mathfrak{M}}^{\infty\perp} \setminus \{0\}$, we have the estimation:

$$\mathrm{P}_t(\mathfrak{M}, \ldots, \mathfrak{M}; |\Psi\rangle) \leq p_{\mathfrak{M}}^{n_{\mathfrak{M}} \lfloor t/n_{\mathfrak{M}} \rfloor} \xrightarrow{t \to \infty} 0$$

§ It is easy to show that, for any quantum state $|\Psi\rangle \in \mathcal{H}$ orthogonal to $\mathcal{H}_{\mathfrak{M}}^{\infty}$ and for any mental state $\mathfrak{M}' \in \mathcal{M}$, $\widehat{U}_{\mathfrak{M}'} |\Psi\rangle$ is still orthogonal to $\mathcal{H}_{\mathfrak{M}}^{\infty}$. For any mental state $\mathfrak{M}' \neq \mathfrak{M}$, this is trivial because $\widehat{U}_{\mathfrak{M}'} |\Psi\rangle \in \mathcal{H}_{\mathfrak{M}'}$, which is orthogonal to the mental subspace $\mathcal{H}_{\mathfrak{M}}$ and *a fortiori* to its subspace $\mathcal{H}_{\mathfrak{M}}^{\infty}$. We assume from now on $\mathfrak{M}' = \mathfrak{M}$. The decomposition $\widehat{\Pi}_{\mathfrak{M}} = \widehat{\Pi}_{\mathcal{H}_{\mathfrak{M}}^{\infty}} + \widehat{\Pi}_{\mathcal{H}_{\mathfrak{M}}^{\infty\perp}}$ reduces to $\widehat{\Pi}_{\mathfrak{M}} |\Psi\rangle = \widehat{\Pi}_{\mathcal{H}_{\mathfrak{M}}^{\infty\perp}} |\Psi\rangle \in \mathcal{H}_{\mathfrak{M}}^{\infty\perp}$ since $|\Psi\rangle$ is orthogonal to $\mathcal{H}_{\mathfrak{M}}^{\infty}$. $\widehat{U}_\tau$ being unitary, it preserves orthogonality, so that $\widehat{U}_\tau \widehat{\Pi}_{\mathfrak{M}} |\Psi\rangle$ is orthogonal to $\widehat{U}_\tau \mathcal{H}_{\mathfrak{M}}^{\infty} = \mathcal{H}_{\mathfrak{M}}^{\infty}$. The above decomposition of $\widehat{\Pi}_{\mathfrak{M}}$ reduces therefore again to $\widehat{\Pi}_{\mathfrak{M}} \widehat{U}_\tau \widehat{\Pi}_{\mathfrak{M}} |\Psi\rangle = \widehat{\Pi}_{\mathcal{H}_{\mathfrak{M}}^{\infty\perp}} \widehat{U}_\tau \widehat{\Pi}_{\mathfrak{M}} |\Psi\rangle$, which proves that $\widehat{U}_{\mathfrak{M}} |\Psi\rangle \in \mathcal{H}_{\mathfrak{M}}^{\infty\perp}$.

§ For any quantum state $|\Psi\rangle \in \mathcal{H}$, any mental state $\mathfrak{M} \in \mathcal{M}$ and any $n \in \mathbb{N}^*$, $\widehat{U}_{\mathfrak{M}}^n |\Psi\rangle$ can thus be decomposed by linearity into the orthogonal sum of a quantum state $\widehat{U}_\tau^n \widehat{\Pi}_{\mathcal{H}_{\mathfrak{M}}^{\infty}} |\Psi\rangle \in \mathcal{H}_{\mathfrak{M}}^{\infty}$ and of a quantum state $\widehat{U}_{\mathfrak{M}}^n \widehat{\Pi}_{\mathcal{H}_{\mathfrak{M}}^{\infty\perp}} |\Psi\rangle \in \mathcal{H}_{\mathfrak{M}}^{\infty\perp}$. Hence its squared norm can be decomposed into:

$$\langle\Psi| \widehat{U}_{\mathfrak{M}}^{\dagger n} \widehat{U}_{\mathfrak{M}}^n |\Psi\rangle = \langle\Psi| \widehat{\Pi}_{\mathcal{H}_{\mathfrak{M}}^{\infty}} |\Psi\rangle + \langle\Psi| \widehat{\Pi}_{\mathcal{H}_{\mathfrak{M}}^{\infty\perp}} \widehat{U}_{\mathfrak{M}}^{\dagger n} \widehat{U}_{\mathfrak{M}}^n \widehat{\Pi}_{\mathcal{H}_{\mathfrak{M}}^{\infty\perp}} |\Psi\rangle$$

If the quantum state $|\Psi\rangle \in \mathcal{H} \setminus \{0\}$ is orthogonal to $\mathcal{H}_{\mathfrak{M}}^{\infty\perp}$, the second term is zero, so that we have for any $t \in \mathbb{N}^*$:

$$\mathrm{P}_t(\mathfrak{M}, \ldots, \mathfrak{M}; |\Psi\rangle) = \langle\Psi| \widehat{\Pi}_{\mathcal{H}_{\mathfrak{M}}^{\infty}} |\Psi\rangle / \langle\Psi|\Psi\rangle$$

Otherwise, for any quantum state $|\Psi\rangle$ having a non-zero component in $\mathcal{H}_{\mathfrak{M}}^{\infty\perp}$, we can write for any $t \in \mathbb{N}$:

$$\mathrm{P}_t(\mathfrak{M}, \ldots, \mathfrak{M}; |\Psi\rangle) = \langle\Psi| \widehat{\Pi}_{\mathcal{H}_{\mathfrak{M}}^{\infty}} |\Psi\rangle / \langle\Psi|\Psi\rangle$$
$$+ \mathrm{P}_t(\mathfrak{M}, \ldots, \mathfrak{M}; \widehat{\Pi}_{\mathcal{H}_{\mathfrak{M}}^{\infty\perp}} |\Psi\rangle) \langle\Psi| \widehat{\Pi}_{\mathcal{H}_{\mathfrak{M}}^{\infty\perp}} |\Psi\rangle / \langle\Psi|\Psi\rangle$$

so that we have more generally:

$$\forall |\Psi\rangle \in \mathcal{H} \setminus \{0\}, \ \mathrm{P}_t(\mathfrak{M}, \ldots, \mathfrak{M}; |\Psi\rangle) \xrightarrow{t \to \infty} \langle\Psi| \widehat{\Pi}_{\mathcal{H}_{\mathfrak{M}}^{\infty}} |\Psi\rangle / \langle\Psi|\Psi\rangle$$

§ It is now easy to conclude. For any sequence $(\mathfrak{M}_t)$ eventually dying at t_f, as stated above, we have for any initial quantum state $|\Psi_i\rangle \in \mathcal{H} \setminus \{0\}$:

$$\mathrm{P}_t(\mathfrak{M}_0, \ldots, \mathfrak{M}_t; |\Psi_i\rangle) \xrightarrow{t \to \infty}$$
$$\langle\Psi_i| \widehat{\Pi}_{\mathfrak{M}_0} \widehat{U}_\tau^\dagger \cdots \widehat{\Pi}_{\mathfrak{M}_{t_f-1}} \widehat{U}_\tau^\dagger \widehat{\Pi}_{\mathcal{H}_{\mathfrak{M}_{t_f}}^{\infty}} \widehat{U}_\tau \widehat{\Pi}_{\mathfrak{M}_{t_f-1}} \cdots \widehat{U}_\tau \widehat{\Pi}_{\mathfrak{M}_0} |\Psi_i\rangle / \langle\Psi_i|\Psi_i\rangle$$

Now we've already seen that, for any quantum state $|\Psi\rangle$ orthogonal to $\mathcal{H}_{\mathfrak{M}_{t_f}}^{\infty}$, $\widehat{U}_\tau |\Psi\rangle$ is orthogonal to this subspace too. Since $\mathfrak{M}_{t_f-1} \neq \mathfrak{M}_{t_f}$ by hypothesis, the subspaces $\mathcal{H}_{\mathfrak{M}t_f-1}$ and $\mathcal{H}^\infty_{\mathfrak{M}t_f} \subset \mathcal{H}_{\mathfrak{M}t_f}$ are orthogonal to each other, so that we have $\widehat{\Pi}_{\mathcal{H}^\infty_{\mathfrak{M}t_f}} \widehat{U}_\tau \widehat{\Pi}_{\mathfrak{M}t_f-1} = 0$. As a consequence,

$$\forall \; |\Psi_i\rangle \in \mathcal{H} \setminus \{0\}, \; \mathrm{P}_t(\mathfrak{M}_0, \ldots, \mathfrak{M}_t; \; |\Psi_i\rangle) \xrightarrow{t \to \infty} 0$$

Any eventually dying sequence is therefore infinitely improbable, Q. E. D.

REMARK This result is actually a quite intuitive consequence of the finite dimensionality of the Hilbert space $\mathcal{H}$. Precisely, the key argument – the compactness of the unit sphere of $\mathcal{H}^{\infty\perp}_{\mathfrak{M}}$ – requires that these subspaces – the "accessible" part of the mental subspaces – be finite dimensional. This condition being given, the result would still hold if the "inaccessible" part $\mathcal{H}^\infty_{\mathfrak{M}}$ of the mental subspaces were infinite dimensional, and/or if there were an infinite number of possible mental states.

THEOREM No quantum state is mortal.

PROOF This is an immediate consequence of the lemma.

COROLLARY There will be with certainty extra-terrestrial forms of conscious life.

PROOF The mental state $\mathfrak{M}_{t_1}$ we are experiencing right now at time t_1 is presumably not the only mental state we have ever experienced, so we take for granted that there has already been a time $t_0 < t_1$ in the past where a mental state $\mathfrak{M}_{t_0} \neq \mathfrak{M}_{t_1}$ has been experienced. The probability that we keep experiencing the same mental state $\mathfrak{M}_{t_1}$ forever from now on can be expressed as a conditional probability of the form:

$$\lim_{t \to \infty} \mathrm{P}_t(\mathfrak{M}_0, \ldots, \mathfrak{M}_t; \; |\Psi_i\rangle)/\mathrm{P}_{t_1}(\mathfrak{M}_0, \ldots, \mathfrak{M}_{t_1}; \; |\Psi_i\rangle)$$

where the sequence $(\mathfrak{M}_t)$ is eventually dying. As a consequence of the lemma, this probability is zero, hence we will experience with certainty a different mental state in the future. This theorem holds for any time $t > t_1$, too: There will be with certainty a time $t' > t$ such that $\mathfrak{M}_{t'} \neq \mathfrak{M}_t$. So basically, once the experienced mental state has changed for the first time in the history of the universe, it will "keep moving" forever, although it is still possible that there are long periods of "mental inactivity" in between.

This applies in particular to the moment when the Sun, in about five billions of years, will have eventually evolved to a red giant and made the Earth an unsuitable place for any form of conscious life. By then, if there isn't yet any extra-terrestrial form of conscious life, the experienced mental state should be $\mathfrak{M}_\Omega$, i.e. the absence of any mental experience. But this state cannot last forever, as we have just seen. If we call "extra-terrestrial form of conscious life" the material substrate of the mental state that would get experienced next, then there will be some with certainty.

COMMENTARIES Proving such theorems in a book on the foundations of physics could be taken as a provocation, but actually I included them here because they are obvious consequences of the formalism and are independent of the details of the physical interactions. They show clearly enough, I think, both the philosophical potential and the dangers of any well-defined interpretation of Quantum Physics. These theorems are in no way a proof of the reality of life after death or of the existence of UFOs, but they would have good chances to get over-interpreted if they would happen to be vulgarized. Developing well-defined interpretations of Quantum Physics obviously has the potential of addressing philosophical issues which are of interest to the public, and it would be a pity not to investigate them from a naturalistic point of view.

7.6   Reincarnation theorem

Generalizing the soul immortality theorem, we will derive here a general theorem on the recurrence of mental states, that we will call "reincarnation theorem" because of its similarity with usual conceptions of reincarnation. A discussion of its precise meaning follows the proof.

NOTATIONS   Let $\mathcal{M}_0$ be a non empty subset of $\mathcal{M}$ and let us write $\mathcal{H}_{\mathcal{M}_0}$ the corresponding mental subspace, given by:

$$\mathcal{H}_{\mathcal{M}_0} := \overset{\perp}{\underset{\mathfrak{M} \in \mathcal{M}_0}{\bigoplus}} \mathcal{H}_{\mathfrak{M}}$$

Let us define by recurrence, for any $n \in \mathbb{N}$, the operators $\widehat{P}_{\mathcal{M}_0}^{(n)}$ by:

$$\widehat{P}_{\mathcal{M}_0}^{(0)} := \widehat{\Pi}_{\mathcal{M}_0} := \sum_{\mathfrak{M} \in \mathcal{M}_0} \widehat{\Pi}_{\mathfrak{M}}$$

$$\widehat{P}_{\mathcal{M}_0}^{(n+1)} := \sum_{\mathfrak{M} \in \mathcal{M}_0} \widehat{\Pi}_{\mathfrak{M}} \widehat{U}_{\tau}^{\dagger} \widehat{P}_{\mathcal{M}_0}^{(n)} \widehat{U}_{\tau} \widehat{\Pi}_{\mathfrak{M}}$$

Finally, let us define by recurrence, for any $n \in \mathbb{N}$, the subspaces $\mathcal{H}_{\mathcal{M}_0}^{n}$ by:

$$\mathcal{H}_{\mathcal{M}_0}^{0} := \mathcal{H}_{\mathcal{M}_0}$$

$$\mathcal{H}_{\mathcal{M}_0}^{n+1} := \mathcal{H}_{\mathcal{M}_0}^{n} \cap \{ |\Psi\rangle \in \mathcal{H} \mid \forall \mathfrak{M}_0, \ldots, \mathfrak{M}_n \in \mathcal{M}_0, \ \widehat{U}_{\tau} \widehat{\Pi}_{\mathfrak{M}_n} \cdots \widehat{U}_{\tau} \widehat{\Pi}_{\mathfrak{M}_0} |\Psi\rangle \in \mathcal{H}_{\mathcal{M}_0} \}$$

Let us write $\mathcal{H}_{\mathcal{M}_0}^{\infty}$ their intersection, $\mathcal{H}_{\mathcal{M}_0}^{\infty \perp}$ its supplementary subspace in $\mathcal{H}_{\mathcal{M}_0}$, and $\widehat{\Pi}_{\mathcal{H}_{\mathcal{M}_0}^{\infty}}$ the orthogonal projection operator on $\mathcal{H}_{\mathcal{M}_0}^{\infty}$.

REMARKS   For any initial quantum state $|\Psi\rangle \neq 0$, $\langle\Psi| \widehat{P}_{\mathcal{M}_0}^{(n)} |\Psi\rangle / \langle\Psi|\Psi\rangle$ represents the probability that the experienced mental state belongs to $\mathcal{M}_0$ at all times $t = 0, \ldots, n$. The operators $\widehat{P}_{\mathcal{M}_0}^{(n)}$ are hermitian and therefore diagonalizable on an orthogonal basis with real eigenvalues; these eigenvalues being probabilities, they lie between 0 and 1. The eigenspace for the eigenvalue 1 can be easily identified with $\mathcal{H}_{\mathcal{M}_0}^{n}$. The subspace $\mathcal{H}_{\mathcal{M}_0}^{\infty}$ is therefore the set of all initial quantum states for which the experienced mental state belongs with certainty to $\mathcal{M}_0$ at all times.

FIRST LEMMA   For any quantum state $|\Psi\rangle \in \mathcal{H}$ orthogonal to $\mathcal{H}_{\mathcal{M}_0}^{\infty}$, and for all mental states $\mathfrak{M} \in \mathcal{M}$, the quantum states $\widehat{U}_{\tau} |\Psi\rangle$ and $\widehat{\Pi}_{\mathfrak{M}} |\Psi\rangle$ are orthogonal to $\mathcal{H}_{\mathcal{M}_0}^{\infty}$, too.

SECOND LEMMA   The sequence of operators $(\widehat{P}_{\mathcal{M}_0}^{(n)})$ converges towards $\widehat{\Pi}_{\mathcal{H}_{\mathcal{M}_0}^{\infty}}$.

THEOREM   An initial mental state $\mathfrak{M}_i \in \mathcal{M}$ being given, for any initial quantum state $|\Psi\rangle \in \mathcal{H}_{\mathfrak{M}_i} \setminus \{0\}$, the probability that the mental state $\mathfrak{M}_i$ be experienced again in the future equals 1.

PROOF OF THE FIRST LEMMA Let us prove first that the subspace $\mathcal{H}_{\mathcal{M}_0}^\infty$ is identical to its image $\widehat{U}_\tau \mathcal{H}_{\mathcal{M}_0}^\infty$. It is easy to see that, for any $n \in \mathbb{N}$, the subspace $\widehat{U}_\tau \mathcal{H}_{\mathcal{M}_0}^{1+n}$ is included in $\mathcal{H}_{\mathcal{M}_0}^n$. Indeed, any quantum state $|\Psi\rangle \in \mathcal{H}_{\mathcal{M}_0}^{1+n}$ belongs per definition to $\mathcal{H}_{\mathcal{M}_0}$ and verifies, for all mental states $\mathfrak{M}_0, \ldots, \mathfrak{M}_n \in \mathcal{M}_0$ and for any $t \in [\![1, n]\!]$, $\widehat{U}_\tau \widehat{\Pi}_{\mathfrak{M}_t} \cdots \widehat{U}_\tau \widehat{\Pi}_{\mathfrak{M}_0} |\Psi\rangle \in \mathcal{H}_{\mathcal{M}_0}$. Summing up over all $\mathfrak{M}_0 \in \mathcal{M}_0$ yields $\widehat{U}_\tau \widehat{\Pi}_{\mathfrak{M}_t} \cdots \widehat{U}_\tau \widehat{\Pi}_{\mathcal{M}_0} |\Psi\rangle \in \mathcal{H}_{\mathcal{M}_0}$ and, since $|\Psi\rangle \in \mathcal{H}_{\mathcal{M}_0}$, $\widehat{U}_\tau \widehat{\Pi}_{\mathfrak{M}_t} \cdots \widehat{U}_\tau |\Psi\rangle \in \mathcal{H}_{\mathcal{M}_0}$. The quantum state $\widehat{U}_\tau |\Psi\rangle$ belongs also to $\mathcal{H}_{\mathcal{M}_0}^n$. We have therefore $\widehat{U}_\tau \mathcal{H}_{\mathcal{M}_0}^{1+n} \subset \mathcal{H}_{\mathcal{M}_0}^n$. Now $\mathcal{H}_{\mathcal{M}_0}^\infty$ is the intersection of all subspaces $\mathcal{H}_{\mathcal{M}_0}^n$ for $n \in \mathbb{N}$, so its image $\widehat{U}_\tau \mathcal{H}_{\mathcal{M}_0}^\infty$ is included in the intersection of their images $\widehat{U}_\tau \mathcal{H}_{\mathcal{M}_0}^n$, and a fortiori in the intersection of the images $\widehat{U}_\tau \mathcal{H}_{\mathcal{M}_0}^{1+n}$, where $n \in \mathbb{N}$. Since they are themselves included in $\mathcal{H}_{\mathcal{M}_0}^n$, we have $\widehat{U}_\tau \mathcal{H}_{\mathcal{M}_0}^\infty \subset \mathcal{H}_{\mathcal{M}_0}^\infty$. The subspace $\mathcal{H}_{\mathcal{M}_0}^\infty$ being finite dimensional and the endomorphism $\widehat{U}_\tau$ injective, we have also $\dim \widehat{U}_\tau \mathcal{H}_{\mathcal{M}_0}^\infty = \dim \mathcal{H}_{\mathcal{M}_0}^\infty$. We have therefore $\widehat{U}_\tau \mathcal{H}_{\mathcal{M}_0}^\infty = \mathcal{H}_{\mathcal{M}_0}^\infty$.

§ We can now prove the first part of the lemma: For any quantum state $|\Psi\rangle \in \mathcal{H}$ orthogonal to $\mathcal{H}_{\mathcal{M}_0}^\infty$, the quantum state $\widehat{U}_\tau |\Psi\rangle$ is orthogonal to $\mathcal{H}_{\mathcal{M}_0}^\infty$, too. Indeed, the elementary evolution operator $\widehat{U}_\tau$ being unitary, the quantum state $\widehat{U}_\tau |\Psi\rangle$ is orthogonal to $\widehat{U}_\tau \mathcal{H}_{\mathcal{M}_0}^\infty$, which is identical to $\mathcal{H}_{\mathcal{M}_0}^\infty$, as we have just seen.

§ Let us consider now, for any $n \in \mathbb{N}$, a quantum state $|\Psi\rangle \in \mathcal{H}_{\mathcal{M}_0}^n$ and a mental state $\mathfrak{M} \in \mathcal{M}$. We will show by recurrence on n that $\widehat{\Pi}_\mathfrak{M} |\Psi\rangle$ belongs to $\mathcal{H}_{\mathcal{M}_0}^n$, too. If $\mathfrak{M} \in \mathcal{M}_0$, it is clear that $\widehat{\Pi}_\mathfrak{M} |\Psi\rangle$ belongs to $\mathcal{H}_{\mathcal{M}_0}^0$, because $\widehat{\Pi}_\mathfrak{M}$ is a projection operator on $\mathcal{H}_\mathfrak{M}$, which is a subspace of $\mathcal{H}_{\mathcal{M}_0} = \mathcal{H}_{\mathcal{M}_0}^0$. If $\mathfrak{M} \notin \mathcal{M}_0$, $\widehat{\Pi}_\mathfrak{M} |\Psi\rangle$ is zero since $|\Psi\rangle$ belongs to $\mathcal{H}_{\mathcal{M}_0}$ by hypothesis, which is orthogonal to $\mathcal{H}_\mathfrak{M}$. So in all cases, $\widehat{\Pi}_\mathfrak{M} |\Psi\rangle \in \mathcal{H}_{\mathcal{M}_0}^0$. Let us assume that the assertion holds for a given $n \in \mathbb{N}$. To prove it at the next rang $n+1$, it is sufficient to show that, for any mental states $\mathfrak{M}_0, \ldots, \mathfrak{M}_n \in \mathcal{M}_0$, the quantum state $\widehat{U}_\tau \widehat{\Pi}_{\mathfrak{M}_n} \cdots \widehat{U}_\tau \widehat{\Pi}_{\mathfrak{M}_0} \widehat{\Pi}_\mathfrak{M} |\Psi\rangle$ belongs to $\mathcal{H}_{\mathcal{M}_0}$. If $\mathfrak{M}_0 \neq \mathfrak{M}$, this quantum state is zero since $\widehat{\Pi}_{\mathfrak{M}_0}$ and $\widehat{\Pi}_\mathfrak{M}$ are orthogonal projection operators on two orthogonal subspaces $\mathcal{H}_{\mathfrak{M}_0}$ and $\mathcal{H}_\mathfrak{M}$. If $\mathfrak{M}_0 = \mathfrak{M}$, considering that, $\widehat{\Pi}_{\mathfrak{M}_0}$ being a projection operator, $\widehat{\Pi}_{\mathfrak{M}_0}^2 = \widehat{\Pi}_{\mathfrak{M}_0}$, this quantum state belongs to $\mathcal{H}_{\mathcal{M}_0}$ by hypothesis, which proves the assertion for all $n \in \mathbb{N}$ by recurrence. We have therefore $\widehat{\Pi}_\mathfrak{M} \mathcal{H}_{\mathcal{M}_0}^n \subset \mathcal{H}_{\mathcal{M}_0}^n$ for any mental state $\mathfrak{M} \in \mathcal{M}$ and all $n \in \mathbb{N}$.

§ One can see easily then that, for any mental state $\mathfrak{M} \in \mathcal{M}$, the projection $\widehat{\Pi}_\mathfrak{M} \mathcal{H}_{\mathcal{M}_0}^\infty$ of the subspace $\mathcal{H}_{\mathcal{M}_0}^\infty$ is included in $\mathcal{H}_{\mathcal{M}_0}^\infty$. Indeed, since $\mathcal{H}_{\mathcal{M}_0}^\infty$ is the intersection of the subspaces $\mathcal{H}_{\mathcal{M}_0}^n$, its image $\widehat{\Pi}_\mathfrak{M} \mathcal{H}_{\mathcal{M}_0}^\infty$ is included in the intersection of the images $\widehat{\Pi}_\mathfrak{M} \mathcal{H}_{\mathcal{M}_0}^n$, which are themselves included in the subspaces $\mathcal{H}_{\mathcal{M}_0}^n$, as we have just seen. Their intersection is therefore included in the intersection of the subspaces $\mathcal{H}_{\mathcal{M}_0}^n$, which is $\mathcal{H}_{\mathcal{M}_0}^\infty$. We have therefore $\widehat{\Pi}_\mathfrak{M} \mathcal{H}_{\mathcal{M}_0}^\infty \subset \mathcal{H}_{\mathcal{M}_0}^\infty$ for any mental state $\mathfrak{M} \in \mathcal{M}$.

§ This will allow us to prove that the subspace $\mathcal{H}_{\mathcal{M}_0}^\infty$ can be decomposed as the orthogonal sum of its projections $\widehat{\Pi}_\mathfrak{M} \mathcal{H}_{\mathcal{M}_0}^\infty$ for $\mathfrak{M} \in \mathcal{M}_0$. This sum is orthogonal since the projections are included in the orthogonal subspaces $\mathcal{H}_\mathfrak{M}$, respectively. It is obvious that $\mathcal{H}_{\mathcal{M}_0}^\infty$ is included in the sum, since any quantum state $|\Psi\rangle \in \mathcal{H}_{\mathcal{M}_0}^\infty$ belongs in particular to $\mathcal{H}_{\mathcal{M}_0}$, so that $|\Psi\rangle = \widehat{\Pi}_{\mathcal{M}_0} |\Psi\rangle = \sum_{\mathfrak{M} \in \mathcal{M}_0} \widehat{\Pi}_\mathfrak{M} |\Psi\rangle$. The inverse inclusion also holds since, as we have just seen, the projections are subspaces of $\mathcal{H}_{\mathcal{M}_0}^{\infty}$. We have therefore $\mathcal{H}_{\mathcal{M}_0}^{\infty} = \overset{\perp}{\oplus}_{\mathfrak{M} \in \mathcal{M}_0} \widehat{\Pi}_{\mathfrak{M}} \mathcal{H}_{\mathcal{M}_0}^{\infty}$.

§ We can conclude from this result that the subspace $\mathcal{H}_{\mathcal{M}_0}^{\infty\perp}$ can be decomposed too as the orthogonal sum of its projections $\widehat{\Pi}_{\mathfrak{M}} \mathcal{H}_{\mathcal{M}_0}^{\infty\perp}$ for $\mathfrak{M} \in \mathcal{M}_0$. Indeed, the subspace $\mathcal{H}_{\mathcal{M}_0}$ being the orthogonal sum of the mental subspaces $\mathcal{H}_{\mathfrak{M}}$ for $\mathfrak{M} \in \mathcal{M}_0$, and the orthogonal components $\widehat{\Pi}_{\mathfrak{M}} \mathcal{H}_{\mathcal{M}_0}^{\infty}$ of $\mathcal{H}_{\mathcal{M}_0}^{\infty}$ being respectively included in the mental subspaces $\mathcal{H}_{\mathfrak{M}}$, the supplementary subspace of $\mathcal{H}_{\mathcal{M}_0}^{\infty}$ in $\mathcal{H}_{\mathcal{M}_0}$ can be decomposed as the orthogonal sum of the respective supplementary subspaces of $\widehat{\Pi}_{\mathfrak{M}} \mathcal{H}_{\mathcal{M}_0}^{\infty}$ in $\mathcal{H}_{\mathfrak{M}}$ – which are therefore its respective projections on the mental subspaces $\mathcal{H}_{\mathfrak{M}}$. We have therefore $\mathcal{H}_{\mathcal{M}_0}^{\infty\perp} = \overset{\perp}{\oplus}_{\mathfrak{M} \in \mathcal{M}_0} \widehat{\Pi}_{\mathfrak{M}} \mathcal{H}_{\mathcal{M}_0}^{\infty\perp}$.

§ We can now prove the second part of the lemma: For any quantum state $|\Psi\rangle \in \mathcal{H}$ orthogonal to $\mathcal{H}_{\mathcal{M}_0}^{\infty}$, and any mental state $\mathfrak{M} \in \mathcal{M}$, the quantum state $\widehat{\Pi}_{\mathfrak{M}} |\Psi\rangle$ is orthogonal to $\mathcal{H}_{\mathcal{M}_0}^{\infty}$, too. Indeed, the supplementary subspace of $\mathcal{H}_{\mathcal{M}_0}^{\infty}$ is the orthogonal sum of its supplementary subspace in $\mathcal{H}_{\mathcal{M}_0}$, which is $\mathcal{H}_{\mathcal{M}_0}^{\infty\perp} = \overset{\perp}{\oplus}_{\mathfrak{M} \in \mathcal{M}_0} \widehat{\Pi}_{\mathfrak{M}} \mathcal{H}_{\mathcal{M}_0}^{\infty\perp}$, and of the supplementary subspace of $\mathcal{H}_{\mathcal{M}_0}$ itself, which is $\overset{\perp}{\oplus}_{\mathfrak{M} \in \mathcal{M} \backslash \mathcal{M}_0} \mathcal{H}_{\mathfrak{M}}$. So if $\mathfrak{M} \notin \mathcal{M}_0$, the quantum state $\widehat{\Pi}_{\mathfrak{M}} |\Psi\rangle$, which belongs to $\mathcal{H}_{\mathfrak{M}}$, is orthogonal to $\mathcal{H}_{\mathcal{M}_0}$, and if $\mathfrak{M} \in \mathcal{M}_0$, the quantum state $\widehat{\Pi}_{\mathfrak{M}} |\Psi\rangle$ belongs to $\widehat{\Pi}_{\mathfrak{M}} \mathcal{H}_{\mathcal{M}_0}^{\infty\perp}$, which is a subspace of $\mathcal{H}_{\mathcal{M}_0}^{\infty\perp}$ and is therefore orthogonal to $\mathcal{H}_{\mathcal{M}_0}^{\infty}$. For any mental state $\mathfrak{M} \in \mathcal{M}$, the quantum state $\widehat{\Pi}_{\mathfrak{M}} |\Psi\rangle$ is therefore orthogonal to $\mathcal{H}_{\mathcal{M}_0}^{\infty}$, Q. E. D.

PROOF OF THE SECOND LEMMA We will show first that, for any $n \in \mathbb{N}$, the subspace $\mathcal{H}_{\mathcal{M}_0}^n$ is the eigenspace of $\widehat{P}_{\mathcal{M}_0}^{(n)}$ for the eigenvalue 1. This is obvious for $n = 0$, where we have $\widehat{P}_{\mathcal{M}_0}^{(0)} = \widehat{\Pi}_{\mathcal{M}_0}$ and $\mathcal{H}_{\mathcal{M}_0}^0 = \mathcal{H}_{\mathcal{M}_0}$. Let us prove first, by recurrence on $n \in \mathbb{N}$, that the subspace $\mathcal{H}_{\mathcal{M}_0}^n$ is included in the eigenspace of $\widehat{P}_{\mathcal{M}_0}^{(n)}$ for the eigenvalue 1. As we've just seen, this assertion holds for $n = 0$. Let us suppose that the assertion is proved for a given $n \in \mathbb{N}$, and let us consider a quantum state $|\Psi\rangle \in \mathcal{H}_{\mathcal{M}_0}^{n+1}$. We have per definition:

$$\widehat{P}_{\mathcal{M}_0}^{(n+1)} |\Psi\rangle = \sum_{\mathfrak{M}_0 \in \mathcal{M}_n} \cdots \sum_{\mathfrak{M}_n \in \mathcal{M}_0} \widehat{\Pi}_{\mathfrak{M}_0} \widehat{U}_\tau^\dagger \cdots \widehat{\Pi}_{\mathfrak{M}_n} \widehat{U}_\tau^\dagger \widehat{\Pi}_{\mathcal{M}_0} \widehat{U}_\tau \widehat{\Pi}_{\mathfrak{M}_n} \cdots \widehat{U}_\tau \widehat{\Pi}_{\mathfrak{M}_0} |\Psi\rangle$$

Since $|\Psi\rangle \in \mathcal{H}_{\mathcal{M}_0}^{n+1}$, the quantum state $\widehat{U}_\tau \widehat{\Pi}_{\mathfrak{M}_n} \cdots \widehat{U}_\tau \widehat{\Pi}_{\mathfrak{M}_0} |\Psi\rangle$ belongs to $\mathcal{H}_{\mathcal{M}_0}$, and since the elementary evolution operator $\widehat{U}_\tau$ is unitary and $\widehat{\Pi}_{\mathfrak{M}_n}$ is a projection operator, we have $\widehat{\Pi}_{\mathfrak{M}_n} \widehat{U}_\tau^\dagger \widehat{U}_\tau \widehat{\Pi}_{\mathfrak{M}_n} = \widehat{\Pi}_{\mathfrak{M}_n}$, so that:

$$\widehat{P}_{\mathcal{M}_0}^{(n+1)} |\Psi\rangle = \widehat{P}_{\mathcal{M}_0}^{(n)} |\Psi\rangle$$

Now the recurrence hypothesis yields $\widehat{P}_{\mathcal{M}_0}^{(n)} |\Psi\rangle = |\Psi\rangle$, so that $\widehat{P}_{\mathcal{M}_0}^{(n+1)} |\Psi\rangle = |\Psi\rangle$, which proves the recurrence.

§ Let us prove now, for any $n \in \mathbb{N}$, the inverse inclusion of the eigenspace of $\widehat{P}_{\mathcal{M}_0}^{(n)}$ for the eigenvalue 1 in the subspace $\mathcal{H}_{\mathcal{M}_0}^n$. Since $\widehat{P}_{\mathcal{M}_0}^{(n)} |\Psi\rangle = |\Psi\rangle$ implies $\langle\Psi| \widehat{P}_{\mathcal{M}_0}^{(n)} |\Psi\rangle = \langle\Psi|\Psi\rangle$ for any quantum state $|\Psi\rangle \in \mathcal{H}$, it is sufficient to prove, by recurrence on $n \in \mathbb{N}$, that $\langle\Psi| \widehat{P}_{\mathcal{M}_0}^{(n)} |\Psi\rangle = \langle\Psi|\Psi\rangle$ implies $|\Psi\rangle \in \mathcal{H}_{\mathcal{M}_0}^n$ for any quantum state $|\Psi\rangle \in \mathcal{H}$. We have already proved this for $n = 0$, so let us assume that the assertion holds for a given $n \in \mathbb{N}$. We have per definition:

$$\langle\Psi|\ \widehat{P}_{\mathcal{M}_0}^{(n+1)}\ |\Psi\rangle \quad = \quad \sum_{\mathfrak{M}_0 \in \mathcal{M}_0} \cdots \sum_{\mathfrak{M}_n \in \mathcal{M}_0} \left\|\widehat{\Pi}_{\mathcal{M}_0} \widehat{U}_\tau \widehat{\Pi}_{\mathfrak{M}_n} \cdots \widehat{U}_\tau \widehat{\Pi}_{\mathfrak{M}_0}\ |\Psi\rangle\right\|^2$$

$$= \quad \sum_{\mathfrak{M}_0 \in \mathcal{M}_0} \cdots \sum_{\mathfrak{M}_n \in \mathcal{M}_0} \sum_{\mathfrak{M} \in \mathcal{M}_0} \left\|\widehat{U}_\tau \widehat{\Pi}_{\mathfrak{M}} \widehat{U}_\tau \widehat{\Pi}_{\mathfrak{M}_n} \cdots \widehat{U}_\tau \widehat{\Pi}_{\mathfrak{M}_0}\ |\Psi\rangle\right\|^2$$

Now for any quantum state $|\Phi\rangle \in \mathcal{H}$, $\sum_{\mathfrak{M} \in \mathcal{M}_0} \|\widehat{U}_\tau \widehat{\Pi}_{\mathfrak{M}}\ |\Phi\rangle\|^2 \leq \langle\Phi|\Phi\rangle$, the case of an equality happening if and only if $|\Phi\rangle \in \mathcal{H}_{\mathcal{M}_0}$. The sequence $(\langle\Psi|\ \widehat{P}_{\mathcal{M}_0}^{(n)}\ |\Psi\rangle)$ is therefore decreasing, with an initial value $\langle\Psi|\ \widehat{P}_{\mathcal{M}_0}^{(0)}\ |\Psi\rangle = \langle\Psi|\ \widehat{\Pi}_{\mathcal{M}_0}\ |\Psi\rangle$, and we have $\langle\Psi|\ \widehat{P}_{\mathcal{M}_0}^{(n+1)}\ |\Psi\rangle = \langle\Psi|\ \widehat{P}_{\mathcal{M}_0}^{(n)}\ |\Psi\rangle$ if and only if, for all $\mathfrak{M}_0, \ldots, \mathfrak{M}_n \in \mathcal{M}_0$, $\widehat{U}_\tau \widehat{\Pi}_{\mathfrak{M}_n} \cdots \widehat{U}_\tau \widehat{\Pi}_{\mathfrak{M}_0}\ |\Psi\rangle \in \mathcal{H}_{\mathcal{M}_0}$. Let us suppose now that we have $\langle\Psi|\ \widehat{P}_{\mathcal{M}_0}^{(n+1)}\ |\Psi\rangle = \langle\Psi|\Psi\rangle$. This implies, because of the decrease of the sequence and of its initial value, that $|\Psi\rangle \in \mathcal{H}_{\mathcal{M}_0}$ and that $\langle\Psi|\ \widehat{P}_{\mathcal{M}_0}^{(n)}\ |\Psi\rangle = \langle\Psi|\Psi\rangle$. We have thus $\langle\Psi|\ \widehat{P}_{\mathcal{M}_0}^{(n+1)}\ |\Psi\rangle = \langle\Psi|\ \widehat{P}_{\mathcal{M}_0}^{(n)}\ |\Psi\rangle$, which implies, as we have just seen, that for all $\mathfrak{M}_0, \ldots, \mathfrak{M}_n \in \mathcal{M}_0$, $\widehat{U}_\tau \widehat{\Pi}_{\mathfrak{M}_n} \cdots \widehat{U}_\tau \widehat{\Pi}_{\mathfrak{M}_0}\ |\Psi\rangle \in \mathcal{H}_{\mathcal{M}_0}$. The recurrence hypothesis yielding $|\Psi\rangle \in \mathcal{H}_{\mathcal{M}_0}^n$, we have therefore $|\Psi\rangle \in \mathcal{H}_{\mathcal{M}_0}^{n+1}$, which proves the recurrence. The eigenspace of $\widehat{P}_{\mathcal{M}_0}^{(n)}$ for the eigenvalue 1 is therefore $\mathcal{H}_{\mathcal{M}_0}^n$ for all $n \in \mathbb{N}$.

§ Now the subspace $\mathcal{H}_{\mathcal{M}_0}$ being finite dimensional, there must exist a finite number of distinct nested subspaces $\mathcal{H}_{\mathcal{M}_0}^n$, so that there exists an integer $n_{\mathcal{M}_0}$ such that:

$$n_{\mathcal{M}_0} := \min\{n \in \mathbb{N} \mid \mathcal{H}_{\mathcal{M}_0}^n = \mathcal{H}_{\mathcal{M}_0}^\infty\}$$

The eigenspace of $\widehat{P}_{\mathcal{M}_0}^{(n)}$ for the eigenvalue 1 is therefore $\mathcal{H}_{\mathcal{M}_0}^\infty$ for any $n \geq n_{\mathcal{M}_0}$. In the special case where $n_{\mathcal{M}_0} = 0$, we have $\widehat{P}_{\mathcal{M}_0}^{(n)} = \widehat{\Pi}_{\mathcal{H}_{\mathcal{M}_0}^\infty}$ for all $n \in \mathbb{N}$, so that the sequence $(\widehat{P}_{\mathcal{M}_0}^{(n)})$ trivially converges towards $\widehat{\Pi}_{\mathcal{H}_{\mathcal{M}_0}^\infty}$. Let us assume from now on that we have $n_{\mathcal{M}_0} \geq 1$. Since $\widehat{P}_{\mathcal{M}_0}^{(n)}$ induces an hermitian operator on $\mathcal{H}_{\mathcal{M}_0}$, it induces also, for $n \geq n_{\mathcal{M}_0}$, an hermitian operator on the supplementary subspace $\mathcal{H}_{\mathcal{M}_0}^{\infty\perp}$ of its eigenspace $\mathcal{H}_{\mathcal{M}_0}^\infty$. It is diagonalizable on an orthogonal basis of $\mathcal{H}_{\mathcal{M}_0}^{\infty\perp}$, with a finite number of real eigenvalues, which are all lying in the interval $[0, 1[$. There exists therefore a number $p_{\mathcal{M}_0} \in [0, 1[$ such that $p_{\mathcal{M}_0}^{n_{\mathcal{M}_0}}$ be the greatest eigenvalue of this operator, which can be defined by:

$$p_{\mathcal{M}_0}^{n_{\mathcal{M}_0}} := \max\{\langle\Psi|\ \widehat{P}_{\mathcal{M}_0}^{(n_{\mathcal{M}_0})}\ |\Psi\rangle / \langle\Psi|\Psi\rangle \mid\ |\Psi\rangle \in \mathcal{H}_{\mathcal{M}_0}^{\infty\perp} \setminus \{0\}\}$$

§ For any initial quantum state $|\Psi\rangle \in \mathcal{H}_{\mathcal{M}_0}^{\infty\perp}$, $\langle\Psi|\ \widehat{P}_{\mathcal{M}_0}^{(n_{\mathcal{M}_0})}\ |\Psi\rangle / \langle\Psi|\Psi\rangle$ represents the probability that the quantum state be in $\mathcal{H}_{\mathcal{M}_0}$ at all times $0, \ldots, n_{\mathcal{M}_0}$. Now, as we have seen in the first lemma, the quantum state at time $n_{\mathcal{M}_0}$ will still be orthogonal to $\mathcal{H}_{\mathcal{M}_0}^\infty$, since it is of the form $\widehat{\Pi}_{\mathfrak{M}_{n_{\mathcal{M}_0}}} \widehat{U}_\tau \cdots \widehat{U}_\tau \widehat{\Pi}_{\mathfrak{M}_0}\ |\Psi\rangle$ for a given sequence of mental states $\mathfrak{M}_0, \ldots, \mathfrak{M}_{n_{\mathcal{M}_0}} \in \mathcal{M}_0$. It belongs therefore to $\mathcal{H}_{\mathcal{M}_0}^{\infty\perp}$, so that the probability that it stays in $\mathcal{H}_{\mathcal{M}_0}$ for another $n_{\mathcal{M}_0}$ time steps is at most

$p_{\mathcal{M}_0}^{n_{\mathcal{M}_0}}$, too. We have therefore:

$$\langle \Psi | \ \widehat{P}_{\mathcal{M}_0}^{(2n_{\mathcal{M}_0})} \ | \Psi \rangle \, / \, \langle \Psi | \Psi \rangle \le p_{\mathcal{M}_0}^{2n_{\mathcal{M}_0}}$$

As we have seen, $\langle \Psi | \ \widehat{P}_{\mathcal{M}_0}^{(t)} \ | \Psi \rangle \, / \, \langle \Psi | \Psi \rangle$ is a decreasing function of t, so that we have more generally, for any $t \in \mathbb{N}$:

$$\langle \Psi | \ \widehat{P}_{\mathcal{M}_0}^{(t)} \ | \Psi \rangle \, / \, \langle \Psi | \Psi \rangle \le p_{\mathcal{M}_0}^{n_{\mathcal{M}_0} \lfloor t/n_{\mathcal{M}_0} \rfloor}$$

where $\lfloor \cdot \rfloor$ denotes the floor function.

§ Let $(|\Psi_i\rangle)$ be an orthonormal basis of the subspace $\mathcal{H}_{\mathcal{M}_0}^{\infty \perp}$. The partial trace of the operator $\widehat{P}_{\mathcal{M}_0}^{(t)}$ on this subspace is given, for any $t \in \mathbb{N}$, by:

$$\mathrm{Tr}_{\mathcal{H}_{\mathcal{M}_0}^{\infty \perp}} \widehat{P}_{\mathcal{M}_0}^{(t)} = \sum_i \langle \Psi_i | \ \widehat{P}_{\mathcal{M}_0}^{(t)} \ | \Psi_i \rangle$$

so that we have:

$$\mathrm{Tr}_{\mathcal{H}_{\mathcal{M}_0}^{\infty \perp}} \widehat{P}_{\mathcal{M}_0}^{(t)} \le p_{\mathcal{M}_0}^{n_{\mathcal{M}_0} \lfloor t/n_{\mathcal{M}_0} \rfloor} \dim \mathcal{H}_{\mathcal{M}_0}^{\infty \perp} \xrightarrow{t \to \infty} 0$$

Now this partial trace is the sum of the eigenvalues multiplied by the dimension of the respective eigenspace, and all these eigenvalues are positive. The greatest eigenvalue of $\widehat{P}_{\mathcal{M}_0}^{(t)}$ on $\mathcal{H}_{\mathcal{M}_0}^{\infty \perp}$ converges therefore towards 0, too, which proves that the sequence of operators $(\widehat{P}_{\mathcal{M}_0}^{(t)})$ converges towards 0 on this subspace.

§ We have already seen the eigenspace of $\widehat{P}_{\mathcal{M}_0}^{(t)}$ for the eigenvalue 1 is $\mathcal{H}_{\mathcal{M}_0}^{\infty}$ for any $t \ge n_{\mathcal{M}_0}$, and that the operator induced on $\mathcal{H}_{\mathcal{M}_0}^{\infty \perp}$ converges towards 0. Furthermore, for any quantum state $|\Psi\rangle$ orthogonal to the subspace $\mathcal{H}_{\mathcal{M}_0}$, we have $\widehat{P}_{\mathcal{M}_0}^{(t)} \ | \Psi \rangle = 0$. The sequence of operators $(\widehat{P}_{\mathcal{M}_0}^{(t)})$ converges therefore towards $\widehat{\Pi}_{\mathcal{H}_{\mathcal{M}_0}^{\infty}}$, Q. E. D.

PROOF OF THE THEOREM An initial mental state $\mathfrak{M}_i \in \mathcal{M}$ and an initial quantum state $|\Psi\rangle \in \mathcal{H}_{\mathfrak{M}_i} \setminus \{0\}$ being given, the probability p_1 that the mental state $\mathfrak{M}_i$ be experienced again at the next time step is given by:

$$p_1 - \langle \Psi | \ \widehat{U}_\tau^\dagger \widehat{\Pi}_{\mathfrak{M}_i} \widehat{U}_\tau \ | \Psi \rangle \, / \, \langle \Psi | \Psi \rangle$$

If the mental state experienced at the next time step is different from $\mathfrak{M}_i$, which happens with a probability $1 - p_1$, then this mental state belongs to $\mathcal{M}_0 := \mathcal{M} \setminus \{\mathfrak{M}_i\}$. More generally, for any $t \in \mathbb{N}^*$, the probability p_t that the mental state $\mathfrak{M}_i$ have been experienced again at any time $t' \le t$ is given by:

$$p_t = 1 - \langle \Psi | \ \widehat{U}_\tau^\dagger \widehat{P}_{\mathcal{M}_0}^{(t-1)} \widehat{U}_\tau \ | \Psi \rangle \, / \, \langle \Psi | \Psi \rangle$$

The second lemma yields:

$$p_t \xrightarrow{t \to \infty} 1 - \langle \Psi | \ \widehat{U}_\tau^\dagger \widehat{\Pi}_{\mathcal{H}_{\mathcal{M}_0}^{\infty}} \widehat{U}_\tau \ | \Psi \rangle \, / \, \langle \Psi | \Psi \rangle$$

and since the initial quantum state $|\Psi\rangle$, belonging to $\mathcal{H}_{\mathfrak{M}_i}$, is orthogonal to $\mathcal{H}_{\mathcal{M}_0}^{\infty}$, the first lemma yields:

$$p_t \xrightarrow{t \to \infty} 1$$

The initial mental state $\mathfrak{M}_i$ will therefore be experienced again in the future with certainty, Q. E. D.

REMARK This proof of the reincarnation theorem relies on the hypothesis that the Hilbert space be finite dimensional. This assumption could be relaxed a little and reduced to the hypothesis that the set of all mental states accessible from the initial quantum state – that it, the last mental states $\mathfrak{M}_n$ of sequences $\mathfrak{M}_0, \ldots, \mathfrak{M}_n$ such that $\widehat{\Pi}_{\mathfrak{M}_n} \widehat{U}_\tau \cdots \widehat{\Pi}_{\mathfrak{M}_0} \widehat{U}_\tau \, |\Psi\rangle \neq 0$ – be finite, and the corresponding mental subspaces finite dimensional.

COMMENTARIES Once again, we are addressing here philosophical, or rather spiritual questions which could not even be expressed in the frame of any scientific theory so far. The reincarnation theorem has been derived with the greatest possible formal rigor, so that it is firmly established upon its foundations, which are compatible with everything we know about quantum physics today. So rejecting without any further discussion the notion of "reincarnation" considered in this theorem – the recurrence of mental states – in no scientifically valid option any more. But it is important in the first place to clarify what the theorem exactly means in order to avoid misunderstandings. In a monist framework, contrary to the dualist conceptions of most religions supporting the idea of reincarnation, there is no mental entity like a soul which could be reincarnated in another body after the death of the last one. The only thing that can be "incarnated" are mental states, i.e. the totality of all subjective experiences taking place at a given instant. And a mental state is being experienced whenever the quantum state, representing the state of affairs in the whole universe at the material level, becomes projected into the corresponding mental subspace, in which there is exactly one brain presenting the corresponding activity pattern for each subjective experience in the mental state. This projection, or "collapse", is a stochastic process following the Born rule, which has been validated uncountable times by all quantum experiments. Now the theorem states that, once a mental state has been experienced, the probability that, in the natural evolution of the physical state via physical interactions and collapse processes, this mental state be never experienced again, vanishes in the limit of an infinitely long time. So if we take for granted the fact that the physical state will keep evolving forever the way it does today, – and in particular if there is no sudden and unpredictable "end of the world" standing before us, – it is certain that a mental state that has been experienced once will be experienced again in the future. This is all the reincarnation theorem is stating; it is a mere recurrence theorem for mental states, quite similar in this respect to the Poincaré recurrence theorem for the state of classical systems in Hamiltonian dynamics. But it is not a recurrence theorem for quantum states: The same mental state can be experienced in an infinite number of different quantum states belonging to the corresponding mental subspace, and even quantum states that differ very slightly from another can yield to very different evolutions on the long term, as we know from quantum chaos theory.

Now it follows trivially from the reincarnation theorem that every single subjective experience, once it has taken place for the first time, will be repeated again and again an infinite number of times. Every single instant of your own mental life, in particular, will be experienced again and again in an infinite number of lives. This is the point where we are getting very near to usual notions of reincarnation. If you identify yourself with your subjective experiences, then you can say that you will be reincarnated in an infinite number of lives. These lives might be very different from another; not all of the subjective experiences of your current life must take place again in each "reincarnation", nor must they occur in the same order, and your subjective experiences might even "reincarnate" in several lives taking place simultaneously (which approaches the notion of "avatars"). Now this notion of reincarnation has nothing supernatural; it happens, so to say, at random, whenever the physical evolution of the universe produces again a brain with an activity pattern corresponding to a subjective experience that has already taken place before. It differs in this regard from the Buddhist conception of reincarnation, for instance, where the soul continues after death its journey on Earth by reincarnating in the body of another, possibly very different living being. Actually, an important part of your subjective experiences probably won't be "reincarnated" on this Earth, because they contain representations of contingent elements of reality – like the last breaking news, technological artifacts, mode accessories – that most probably won't occur again in the culture history of Manhood. But still they will be "incarnated" again – on another planet quite similar to Earth. And even your vision of the constellations in the night sky will be "reincarnated" once – in another galaxy quite similar to the Milky Way. Now how is this possible? The latest astronomical observations, interpreted in the frame of the Λ-Cold Dark Matter model (a refinement of the Friedmann–Lemaître–Robertson–Walker general relativistic, homogeneous, isotropic model of the universe [11]), suggest that the universe is going to keep expanding at a constant rate, the subsequent dilution of matter preventing on the long term the formation of new galaxies and stars, which will eventually all get extincted. This is the so-called "cold death" scenario in cosmology. Why do we seem to escape this scenario in lattice quantum field theory? This is definitely not related to the details of the physical interactions, and in particular of the choice of the graviton model we could use, since the reincarnation theorem doesn't depend on the exact form of the elementary evolution operator $\widehat{U}_\tau$. The situation would certainly look different if the elementary evolution operator would change across time, which would probably be the case if we computed a semi-classical gravitational background at each time step, for instance. But for now, with a constant elementary evolution operator, the model cannot account for an expansion of space-time; the galaxies, if they are drifting apart consequently to a "big bang" event, would meet again after having traveled through half of the universe because of its toroidal character. This would be kind of a "big crunch" event (but not due to a contraction of space-time) which would be followed by another "big bang" event that could possibly yield to the formation of another Milky Way and of another Earth on which your subjective experiences could be "reincarnated". This conception is quite similar to Buddhist cosmology, where the world is supposed to cyclically come to existence and dissolve again.

There are several ways of questioning the reality of this "reincarnation" in a purely scientific approach. The reincarnation theorem relies on the finite dimensionality of the Hilbert space of quantum states, and thus on the assumption that the physical space itself is discrete and finite, which has elusive, but in principle measurable consequences. Observing experimentally structures in differential scattering cross-sections revealing the finiteness of the lattice step or the quantization of momentum, or observing patterns in the angular distribution of the cosmic microwave background radiation revealing the toroidal character of space, would provide hints in favor of this model and thus of the reincarnation theorem, while their non-observation would put experimental constraints on its parameter space. On the other hand, in a theoretical approach, fulfilling the program of constructive quantum field theory – constructing a well-defined theory of interacting fields, compatible with the Standard Model, on the four dimensional Minkowski space-time – would provide an alternative to our model where the Hilbert space presumably wouldn't be finite dimensional, so that the reincarnation theorem probably wouldn't hold. Finally, on a conceptual level, developing an alternative model of the mental world or of the mind-body relationship within the frame of quantum theory would be a promising option too, since this is a completely new field of science. There are probably plenty of interesting models to investigate once one has accepted to put subjective experience in an equation in the first place.

Part IV

Physical interactions

Chapter 8

Quantum Electrodynamics

In this chapter, we will define the interaction Hamiltonian of Quantum Electrodynamics (QED), describing the photon mediated electromagnetic interaction between electrically charged fermions, and we will derive the composition of the corresponding dressed particles.

8.1   Electric charge operator

On each point n of space, the electric charge operator is defined by:

$$\widehat{Q}_{\boldsymbol{n}} := \mathrm{e} \sum_{\phi, \lambda', \lambda} Q_\phi \left(\widehat{\overline{\psi^\phi}}_{\boldsymbol{n}, \lambda'} + \widehat{\overline{\psi^\phi}}_{\boldsymbol{n}, \lambda'} \right) \gamma^0 \left(\widehat{\psi^\phi}_{\boldsymbol{n}, \lambda} + \widehat{\overline{\overline{\psi^\phi}}}_{\boldsymbol{n}, \lambda} \right)$$

where e is the elementary electric charge (opposite electric charge of a bare electron) and Q_ϕ the electric charge number of fermions of type ϕ: $Q_\phi = 0$ for the neutrinos $\phi \in \{\nu_e, \nu_\mu, \nu_\tau\}$, $Q_\phi = -1$ for the charged leptons $\phi \in \{e, \mu, \tau\}$, $Q_\phi = \frac{2}{3}$ for the quarks $\phi \in \{u, c, t\}$ and $Q_\phi = -\frac{1}{3}$ for the quarks $\phi \in \{d, s, b\}$. In this expression, the creation and annihilation spinor operators are defined as in appendix C.3 and the Dirac matrices as in appendix B.2.

The anti-particle creation and annihilation spinor operators in this expression yield to a uniformly distributed mean electric charge in the vacuum state, given by:

$$\langle \Omega | \ \widehat{Q}_{\boldsymbol{n}} \ | \Omega \rangle = \mathrm{e} \sum_{\phi, \lambda} Q_\phi = -4\mathrm{e}$$

and called 'zero-point electric charge'. Since this charge distribution is uniform, it doesn't have any contribution to the interaction Hamiltonian as defined in section 8.5.

8.2   Electric current operator

On each point $\boldsymbol{n}$ of space, the electric current operator is defined by:

$$\widehat{\boldsymbol{J}}_{\boldsymbol{n}} := \text{ec} \sum_{\phi,\lambda',\lambda} Q_\phi \left(\widehat{\overline{\psi^\phi}}_{\boldsymbol{n},\lambda'} + \widehat{\overline{\psi^\phi}}_{\boldsymbol{n},\lambda'} \right) \gamma \left(\widehat{\psi^\phi}_{\boldsymbol{n},\lambda} + \widehat{\overline{\psi^\phi}}_{\boldsymbol{n},\lambda} \right)$$

where the summation goes over all fermions ϕ. In this expression too, the creation and annihilation spinor operators are defined as in appendix C.3 and the Dirac matrices as in appendix B.2.

8.3   Electric potential operator

On each point $\boldsymbol{n}$ of space, the electric potential operator is defined by:

$$\widehat{V}_{\boldsymbol{n}} := \sum_{\boldsymbol{n}'} \frac{\widehat{Q}_{\boldsymbol{n}'}}{8\pi\varepsilon_0 a} (1 + 2\mathrm{N})^{-3} \sum_{\boldsymbol{q}_\gamma \neq \boldsymbol{0}} \frac{\exp\left(\mathrm{i}2\pi \boldsymbol{q}_\gamma \cdot (\boldsymbol{n} - \boldsymbol{n}') \right)}{\pi q_\gamma^2}$$

where ε_0 is the permittivity of the bare vacuum. Its constant Fourier component has been set to 0 (which is consistent with the Coulomb gauge condition used in appendix C.1), so that the zero-point electric charge in section 8.1 doesn't have any contribution to the interaction Hamiltonian as defined in section 8.5.

8.4   Magnetic potential operator

On each point $\boldsymbol{n}$ of space, the magnetic potential operator is defined by:

$$\widehat{\boldsymbol{A}}_{\boldsymbol{n}} := \sum_{\lambda_\gamma} \widehat{\psi^\gamma}^\dagger_{\boldsymbol{n},\lambda_\gamma} + \widehat{\psi^\gamma}_{\boldsymbol{n},\lambda_\gamma}$$

In this expression, the creation and annihilation spinor operators are defined as in appendix C.1.

8.5   QED interaction Hamiltonian

The interaction Hamiltonian of QED is defined by:

$$\widehat{\mathrm{H}}'_{QED} := \sum_{\boldsymbol{n}} \widehat{\boldsymbol{J}}_{\boldsymbol{n}} \cdot \widehat{\boldsymbol{A}}_{\boldsymbol{n}} + \widehat{Q}_{\boldsymbol{n}} \widehat{V}_{\boldsymbol{n}}$$

Its development on the plane waves basis is given by:

$$\sum_n \widehat{\boldsymbol{J}}_n \cdot \widehat{\boldsymbol{A}}_n = \sqrt{\frac{e^2 hc}{8\pi^2 \varepsilon_0 a^2}} (1+2N)^{-3/2} \sum_{\phi,\boldsymbol{q},\lambda',\lambda} Q_\phi \sum_{\boldsymbol{q}_\gamma \neq 0, \lambda_\gamma} q_\gamma^{-1/2}$$

$$\left[\left(\widehat{\overline{\psi}^\phi}_{\boldsymbol{q}-\boldsymbol{q}_\gamma,\lambda'} + \widehat{\psi^{\overline{\phi}}}_{-\boldsymbol{q}+\boldsymbol{q}_\gamma,\lambda'} \right) \gamma \left(\widehat{\psi^\phi}_{\boldsymbol{q},\lambda} + \widehat{\overline{\psi}^{\overline{\phi}}}_{-\boldsymbol{q},\lambda} \right) \cdot \right.$$

$$\varepsilon^*_{\boldsymbol{q}_\gamma,\lambda_\gamma} \widehat{a^\gamma}^\dagger_{\boldsymbol{q}_\gamma,\lambda_\gamma} \sqrt{1 + \widehat{N^\gamma}_{\boldsymbol{q}_\gamma,\lambda_\gamma}} +$$

$$\left(\widehat{\overline{\psi}^\phi}_{\boldsymbol{q}+\boldsymbol{q}_\gamma,\lambda'} + \widehat{\psi^{\overline{\phi}}}_{-\boldsymbol{q}-\boldsymbol{q}_\gamma,\lambda'} \right) \gamma \left(\widehat{\psi^\phi}_{\boldsymbol{q},\lambda} + \widehat{\overline{\psi}^{\overline{\phi}}}_{-\boldsymbol{q},\lambda} \right) \cdot$$

$$\left. \varepsilon_{\boldsymbol{q}_\gamma,\lambda_\gamma} \widehat{a^\gamma}_{\boldsymbol{q}_\gamma,\lambda_\gamma} \sqrt{\widehat{N^\gamma}_{\boldsymbol{q}_\gamma,\lambda_\gamma}} \right]$$

$$\sum_n \widehat{Q}_n \widehat{V}_n = \frac{e^2}{8\pi^2 \varepsilon_0 a} (1+2N)^{-3} \sum_{\phi,\boldsymbol{q},\lambda',\lambda} Q_\phi \sum_{\phi_0,\boldsymbol{q}_0,\lambda'_0,\lambda_0} Q_{\phi_0} \sum_{\boldsymbol{q}_\gamma \neq 0} q_\gamma^{-2}$$

$$\left(\widehat{\overline{\psi}^\phi}_{\boldsymbol{q}+\boldsymbol{q}_\gamma,\lambda'} + \widehat{\psi^{\overline{\phi}}}_{-\boldsymbol{q}-\boldsymbol{q}_\gamma,\lambda'} \right) \gamma^0 \left(\widehat{\psi^\phi}_{\boldsymbol{q},\lambda} + \widehat{\overline{\psi}^{\overline{\phi}}}_{-\boldsymbol{q},\lambda} \right)$$

$$\left(\widehat{\overline{\psi}^{\phi_0}}_{\boldsymbol{q}_0-\boldsymbol{q}_\gamma,\lambda'_0} + \widehat{\psi^{\overline{\phi_0}}}_{-\boldsymbol{q}_0+\boldsymbol{q}_\gamma,\lambda'_0} \right) \gamma^0 \left(\widehat{\psi^{\phi_0}}_{\boldsymbol{q}_0,\lambda_0} + \widehat{\overline{\psi}^{\overline{\phi_0}}}_{-\boldsymbol{q}_0,\lambda_0} \right)$$

8.6   Dressed states

As a consequence of the electromagnetic interaction between the photon and the charged fermion fields, an excitation of a single particle field like $\left| N^\phi_{\boldsymbol{q},\lambda} \right\rangle$ is unstable and is also a poor model for observed particles. In fact, these particles are always being observed "dressed", *i.e.* forming a particle complex together with excitations of the other fields. As a consequence, the "bare" rest mass, electric charge and magnetic moment of these particles, as they appear in the QED model, do not correspond to the values observed by dressed particles. These renormalized values as well as the composition of dressed particles can be calculated in the frame of QED as a function of the bare values, which can also be indirectly determined experimentally.

We consider a bare state of the form $\left| (N^\phi_{\boldsymbol{q},\lambda})_0 \right\rangle$ and will derive the corresponding dressed state $|\Psi\rangle$ as eigenstate of $\widehat{H}_0 + \widehat{H}'$ for an eigenvalue E to be determined. Assuming the bare and dressed states aren't orthogonal to each other, we write the latter as:

$$|\Psi\rangle := \widetilde{\Psi}\left((N^\phi_{\boldsymbol{q},\lambda})_0 \right) \sum_{(N^\phi_{\boldsymbol{q},\lambda})} \widetilde{\Phi}_0 \left((N^\phi_{\boldsymbol{q},\lambda}) \right) \left| (N^\phi_{\boldsymbol{q},\lambda}) \right\rangle$$

using unnormalized coefficients $\widetilde{\Phi}_0$ verifying the condition:

$$\widetilde{\Phi}_0 \left((N^\phi_{\boldsymbol{q},\lambda})_0 \right) = 1$$

The eigenvalue equation, projected on $\left|(N^\phi_{q,\lambda})_0\right\rangle$ resp. on another plane wave state $\left|(N^\phi_{q,\lambda})_1\right\rangle$, reads:

$$H'_{0,0} + \sum_{(N^\phi_{q,\lambda})_2 \neq (N^\phi_{q,\lambda})_0} H'_{0,2}\widetilde{\Phi}_{2,0} = E - E_0$$

$$H'_{1,0} + \sum_{(N^\phi_{q,\lambda})_2 \neq (N^\phi_{q,\lambda})_0} H'_{1,2}\widetilde{\Phi}_{2,0} = (E - E_1)\widetilde{\Phi}_{1,0}$$

where we use the shorthand notations:

$$H'_{b,a} := \left\langle (N^\phi_{q,\lambda})_b\right| \widehat{H}' \left|(N^\phi_{q,\lambda})_a\right\rangle$$

$$E_a := \left\langle (N^\phi_{q,\lambda})_a\right| \widehat{H}_0 \left|(N^\phi_{q,\lambda})_a\right\rangle$$

$$\widetilde{\Phi}_{a,0} := \widetilde{\Phi}_0\left((N^\phi_{q,\lambda})_a\right)$$

To solve this equation iteratively, we develop $\widehat{H}'$, $\widetilde{\Phi}_0$ and E as power series in the elementary electric charge e:

$$\widehat{H}' := \widehat{H}'^{(1)} + \widehat{H}'^{(2)}$$

$$\widetilde{\Phi}_0 := \sum_{n=0}^{\infty} \widetilde{\Phi}_0^{(n)}$$

$$E := \sum_{n=0}^{\infty} E^{(n)}$$

where we take:

$$\widehat{H}'^{(1)} := \sum_n \widehat{J}_n \cdot \widehat{A}_n$$

$$\widehat{H}'^{(2)} := \sum_n \widehat{Q}_n \widehat{V}_n$$

$$\widetilde{\Phi}_{a,0}^{(0)} := \delta_{a,0}$$

In the case of bare states which are nondegenerate with respect to $\widehat{H}_0$, i.e. such that $E_2 \neq E_0$ for any $(N^\phi_{q,\lambda})_2 \neq (N^\phi_{q,\lambda})_0$, assuming that the eigenvalue equation should hold to each order separately, we have to the zeroth order:

$$E^{(0)} = E_0$$

to the first order:

$$E^{(1)} = H'^{(1)}_{0,0}$$

$$\widetilde{\Phi}^{(1)}_{1,0} = \frac{H'^{(1)}_{1,0}}{E_0 - E_1}$$

to the second order:

$$E^{(2)} = H_{0,0}'^{(2)} + \sum_{(N_{q,\lambda}^\phi)_2 \neq (N_{q,\lambda}^\phi)_0} \frac{H_{0,2}'^{(1)} H_{2,0}'^{(1)}}{E_0 - E_2}$$

$$\widetilde{\Phi}_{1,0}^{(2)} = \frac{H_{1,0}'^{(2)}}{E_0 - E_1} + \sum_{(N_{q,\lambda}^\phi)_2 \neq (N_{q,\lambda}^\phi)_0} \frac{H_{1,2}'^{(1)} H_{2,0}'^{(1)}}{(E_0 - E_1)(E_0 - E_2)} - \frac{H_{1,0}'^{(1)} H_{0,0}'^{(1)}}{(E_0 - E_1)^2}$$

and to the order $n > 2$:

$$E^{(n)} = \sum_{(N_{q,\lambda}^\phi)_2 \neq (N_{q,\lambda}^\phi)_0} \left(H_{0,2}'^{(1)} \widetilde{\Phi}_{2,0}^{(n-1)} + H_{0,2}'^{(2)} \widetilde{\Phi}_{2,0}^{(n-2)} \right)$$

$$\widetilde{\Phi}_{1,0}^{(n)} = \sum_{(N_{q,\lambda}^\phi)_2 \neq (N_{q,\lambda}^\phi)_0} \frac{H_{1,2}'^{(1)} \widetilde{\Phi}_{2,0}^{(n-1)} + H_{1,2}'^{(2)} \widetilde{\Phi}_{2,0}^{(n-2)}}{E_0 - E_1} - \sum_{m=1}^{n-1} \widetilde{\Phi}_{1,0}^{(m)} \frac{E^{(n-m)}}{E_0 - E_1}$$

8.7 Dressed vacuum

The vacuum state itself isn't stable and would become populated by pair creation processes. Up to the first order, the dressed vacuum is composed of the bare vacuum $|\Omega\rangle$ as well as of states of the form $\left| 1_{q-q_\gamma,\lambda'}^\phi 1_{-q,\lambda}^{\overline{\phi}} 1_{q_\gamma,\lambda_\gamma}^\gamma \right\rangle$, where ϕ is any electrically charged fermion and $q_\gamma \neq 0$. The corresponding unnormalized coefficients are given by:

$$\widetilde{\Phi}^{(1)} = -\sqrt{\frac{e^2}{4\pi\varepsilon_0 hc}} (1+2N)^{-3/2} Q_\phi \frac{u_{q-q_\gamma,\lambda'}^{\phi\dagger} \gamma^0 \gamma u_{-q,\lambda}^{\overline{\phi}} \cdot \varepsilon_{q_\gamma,\lambda_\gamma}^*}{(2\pi q_\gamma)^{1/2} \left(E_{q-q_\gamma}^\phi + E_{-q}^{\overline{\phi}} + E_{q_\gamma}^\gamma \right) a/hc}$$

The corresponding energy is of second order and can be written as:

$$E^{(2)} = (1+2N)^3 \frac{e^2}{4\pi\varepsilon_0 a} \sum_\phi Q_\phi^2 \kappa_\phi^\Omega$$

$$\kappa_\phi^\Omega := (1+2N)^{-6} \sum_{q,q_\gamma \neq 0} \left[\frac{1}{2\pi q_\gamma^2} \sum_{\lambda',\lambda} \left| u_{q-q_\gamma,\lambda'}^{\phi\dagger} u_{-q,\lambda}^{\overline{\phi}} \right|^2 \right.$$

$$\left. - \frac{\sum_{\lambda',\lambda,\lambda_\gamma} \left| u_{q-q_\gamma,\lambda'}^{\phi\dagger} \gamma^0 \gamma u_{-q,\lambda}^{\overline{\phi}} \cdot \varepsilon_{q_\gamma,\lambda_\gamma}^* \right|^2}{2\pi q_\gamma \left(E_{q-q_\gamma}^\phi + E_{-q}^{\overline{\phi}} + E_{q_\gamma}^\gamma \right) a/hc} \right]$$

where the spin summations evaluate to:

$$\sum_{\lambda',\lambda} \left| u_{q-q_\gamma,\lambda'}^{\phi\dagger} u_{-q,\lambda}^{\overline{\phi}} \right|^2 = 1 - \frac{M_\phi^2 + q - q_\gamma \cdot q}{E_{q-q_\gamma}^\phi E_{-q}^{\overline{\phi}} (a/hc)^2}$$

$$\sum_{\lambda',\lambda,\lambda_\gamma} \left| u^{\phi\,\dagger}_{q-q_\gamma,\lambda'} \gamma^0 \gamma u^{\overline{\phi}}_{-q,\lambda} \cdot \varepsilon^*_{q_\gamma,\lambda_\gamma} \right|^2 =$$

$$2 \left(1 + \frac{M_\phi^2 + (q - q_\gamma \cdot q_\gamma)(q \cdot q_\gamma)/q_\gamma^2}{E^\phi_{q-q_\gamma} E^{\overline{\phi}}_{-q}(a/hc)^2} \right)$$

For $m_\phi = 0$ and in the special cases where $q = 0$ or $q - q_\gamma = 0$, they evaluate respectively to 1 and 2. The numerical coefficients κ_ϕ^Ω only depend on the reduced masses of the bare particles and are plotted below:

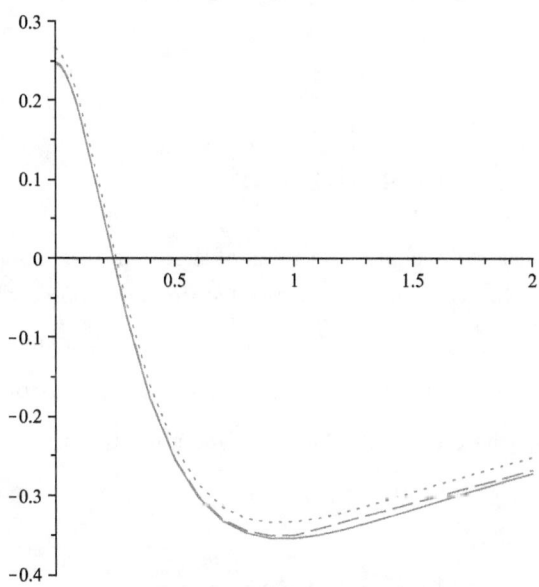

κ_ϕ^Ω as a function of M_ϕ
for N = 1, 2 and 3 (dotted, dashed and solid lines)

In $M_\phi = 0$, we have $\kappa_\phi^\Omega \approx 0.266$, 0.248 and 0.246 for N = 1, 2 and 3 respectively; as $M_\phi \to \infty$, we have $\kappa_\phi^\Omega \to 0^-$. Since the result converges to an integral expression for $N \to \infty$, I shall assume that the coefficients obtained by carrying out the computation for N = 3 are already a good approximation.

This energy is represented by following Feynman diagram:

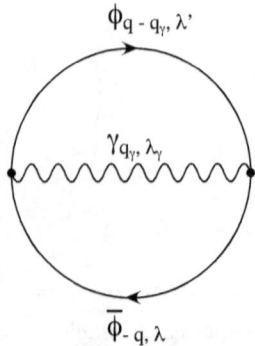

where the Coulomb interaction term is conventionally represented by the case $\lambda_\gamma = 0$. Assuming $\kappa_\phi^\Omega \approx 0.25$ for electrically charged fermions, the energy of the electromagnetically dressed vacuum evaluates up to the second order to:

$$E^{(2)} \approx 0.25(1 + 2\mathrm{N})^3 \frac{14}{3} \frac{e^2}{4\pi\varepsilon_0 a}$$

8.8 Dressed charged fermion

We consider an electrically charged fermion of type f in the bare state $\left| 1^f_{\boldsymbol{q}_f, \lambda_f} \right\rangle$. Up to the first order, the corresponding dressed state is composed of the bare state, of states of the form $\left| 1^f_{\boldsymbol{q}_f, \lambda_f} 1^\phi_{\boldsymbol{q} - \boldsymbol{q}_\gamma, \lambda'} 1^{\overline{\phi}}_{-\boldsymbol{q}, \lambda} 1^\gamma_{\boldsymbol{q}_\gamma, \lambda_\gamma} \right\rangle$, where ϕ is any electrically charged fermion, $\boldsymbol{q}_\gamma \neq \boldsymbol{0}$ and $(\phi, \boldsymbol{q} - \boldsymbol{q}_\gamma, \lambda') \neq (f, \boldsymbol{q}_f, \lambda_f)$, as well as of states of the form $\left| 1^f_{\boldsymbol{q}_f - \boldsymbol{q}_\gamma, \lambda} 1^\gamma_{\boldsymbol{q}_\gamma, \lambda_\gamma} \right\rangle$, where $\boldsymbol{q}_\gamma \neq \boldsymbol{0}$. The corresponding unnormalized coefficients are given by:

$$\widetilde{\Phi}^{(1)} = -\sqrt{\frac{e^2}{4\pi\varepsilon_0 \hbar c}}(1 + 2\mathrm{N})^{-3/2} Q_\phi \frac{u^{\phi\dagger}_{\boldsymbol{q} - \boldsymbol{q}_\gamma, \lambda'} \gamma^0 \gamma u^{\overline{\phi}}_{-\boldsymbol{q}, \lambda} \cdot \varepsilon^*_{\boldsymbol{q}_\gamma, \lambda_\gamma}}{(2\pi q_\gamma)^{1/2} \left(E^\phi_{\boldsymbol{q} - \boldsymbol{q}_\gamma} + E^{\overline{\phi}}_{-\boldsymbol{q}} + E^\gamma_{\boldsymbol{q}_\gamma} \right) a/\hbar c}$$

for states of the form $\left| 1^f_{\boldsymbol{q}_f, \lambda_f} 1^\phi_{\boldsymbol{q} - \boldsymbol{q}_\gamma, \lambda'} 1^{\overline{\phi}}_{-\boldsymbol{q}, \lambda} 1^\gamma_{\boldsymbol{q}_\gamma, \lambda_\gamma} \right\rangle$, and by:

$$\widetilde{\Phi}^{(1)} = -\sqrt{\frac{e^2}{4\pi\varepsilon_0 \hbar c}}(1 + 2\mathrm{N})^{-3/2} Q_f \frac{u^{f\dagger}_{\boldsymbol{q}_f - \boldsymbol{q}_\gamma, \lambda} \gamma^0 \gamma u^f_{\boldsymbol{q}_f, \lambda_f} \cdot \varepsilon^*_{\boldsymbol{q}_\gamma, \lambda_\gamma}}{(2\pi q_\gamma)^{1/2} \left(E^f_{\boldsymbol{q}_f - \boldsymbol{q}_\gamma} + E^\gamma_{\boldsymbol{q}_\gamma} - E^f_{\boldsymbol{q}_f} \right) a/\hbar c}$$

for states of the form $\left| 1^f_{\boldsymbol{q}_f - \boldsymbol{q}_\gamma, \lambda} 1^\gamma_{\boldsymbol{q}_\gamma, \lambda_\gamma} \right\rangle$, respectively.

The corresponding energy is of second order and can be written as:

$$
\begin{aligned}
E^{(2)} \;=\; & E^{(2)}\left(\Omega\right) + \frac{e^2}{4\pi\varepsilon_0 a} Q_f^2 \kappa^f_{\boldsymbol{q}_f,\lambda_f} \\[4pt]
\kappa^f_{\boldsymbol{q}_f,\lambda_f} \;:=\; & (1+2\mathrm{N})^{-3} \sum_{\boldsymbol{q}_\gamma \neq 0} \left[\frac{1}{2\pi q_\gamma^2} \sum_\lambda \left| u^{f\,\dagger}_{\boldsymbol{q}_f-\boldsymbol{q}_\gamma,\lambda} u^{f}_{\boldsymbol{q}_f,\lambda_f} \right|^2 \right. \\[4pt]
& \left. - \frac{\sum_{\lambda,\lambda_\gamma} \left| u^{f\,\dagger}_{\boldsymbol{q}_f-\boldsymbol{q}_\gamma,\lambda} \gamma^0 \gamma u^{f}_{\boldsymbol{q}_f,\lambda_f} \cdot \varepsilon^*_{\boldsymbol{q}_\gamma,\lambda_\gamma} \right|^2}{2\pi q_\gamma \left(E^f_{\boldsymbol{q}_f-\boldsymbol{q}_\gamma} + E^\gamma_{\boldsymbol{q}_\gamma} - E^f_{\boldsymbol{q}_f} \right) a/hc} \right] \\[4pt]
& - (1+2\mathrm{N})^{-3} \sum_{\boldsymbol{q}_\gamma \neq 0} \left[\frac{1}{2\pi q_\gamma^2} \sum_\lambda \left| u^{f\,\dagger}_{\boldsymbol{q}_f,\lambda_f} u^{\overline{f}}_{-\boldsymbol{q}_f-\boldsymbol{q}_\gamma,\lambda} \right|^2 \right. \\[4pt]
& \left. - \frac{\sum_{\lambda,\lambda_\gamma} \left| u^{f\,\dagger}_{\boldsymbol{q}_f,\lambda_f} \gamma^0 \gamma u^{\overline{f}}_{-\boldsymbol{q}_f-\boldsymbol{q}_\gamma,\lambda} \cdot \varepsilon^*_{\boldsymbol{q}_\gamma,\lambda_\gamma} \right|^2}{2\pi q_\gamma \left(E^f_{\boldsymbol{q}_f} + E^{\overline{f}}_{-\boldsymbol{q}_f-\boldsymbol{q}_\gamma} + E^\gamma_{\boldsymbol{q}_\gamma} \right) a/hc} \right]
\end{aligned}
$$

where $E^{(2)}\left(\Omega\right)$ is the second order energy of the dressed vacuum. The spin summations evaluate to:

$$
\sum_\lambda \left| u^{f\,\dagger}_{\boldsymbol{q}_f-\boldsymbol{q}_\gamma,\lambda} u^{f}_{\boldsymbol{q}_f,\lambda_f} \right|^2 = \frac{1}{2}\left(1 + \frac{\mathrm{M}_f^2 + \boldsymbol{q}_f - \boldsymbol{q}_\gamma \cdot \boldsymbol{q}_f}{E^f_{\boldsymbol{q}_f-\boldsymbol{q}_\gamma} E^f_{\boldsymbol{q}_f}(a/hc)^2} \right)
$$

$$
\sum_{\lambda,\lambda_\gamma} \left| u^{f\,\dagger}_{\boldsymbol{q}_f-\boldsymbol{q}_\gamma,\lambda} \gamma^0 \gamma u^{f}_{\boldsymbol{q}_f,\lambda_f} \cdot \varepsilon^*_{\boldsymbol{q}_\gamma,\lambda_\gamma} \right|^2 = 1 - \frac{\mathrm{M}_f^2 + (\boldsymbol{q}_f - \boldsymbol{q}_\gamma \cdot \boldsymbol{q}_\gamma)(\boldsymbol{q}_f \cdot \boldsymbol{q}_\gamma)/q_\gamma^2}{E^f_{\boldsymbol{q}_f-\boldsymbol{q}_\gamma} E^f_{\boldsymbol{q}_f}(a/hc)^2}
$$

$$
\sum_\lambda \left| u^{f\,\dagger}_{\boldsymbol{q}_f,\lambda_f} u^{\overline{f}}_{-\boldsymbol{q}_f-\boldsymbol{q}_\gamma,\lambda} \right|^2 = \frac{1}{2}\left(1 - \frac{\mathrm{M}_f^2 + \boldsymbol{q}_f \cdot \boldsymbol{q}_f + \boldsymbol{q}_\gamma}{E^f_{\boldsymbol{q}_f} E^{\overline{f}}_{-\boldsymbol{q}_f-\boldsymbol{q}_\gamma}(a/hc)^2} \right)
$$

$$
\sum_{\lambda,\lambda_\gamma} \left| u^{f\,\dagger}_{\boldsymbol{q}_f,\lambda_f} \gamma^0 \gamma u^{\overline{f}}_{-\boldsymbol{q}_f-\boldsymbol{q}_\gamma,\lambda} \cdot \varepsilon^*_{\boldsymbol{q}_\gamma,\lambda_\gamma} \right|^2 = 1 + \frac{\mathrm{M}_f^2 + (\boldsymbol{q}_f \cdot \boldsymbol{q}_\gamma)(\boldsymbol{q}_f + \boldsymbol{q}_\gamma \cdot \boldsymbol{q}_\gamma)/q_\gamma^2}{E^f_{\boldsymbol{q}_f} E^{\overline{f}}_{-\boldsymbol{q}_f-\boldsymbol{q}_\gamma}(a/hc)^2}
$$

The vacuum energy diagram is also completed by subtracting following contribution:

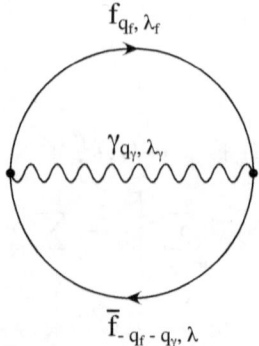

and by adding following self-energy diagram:

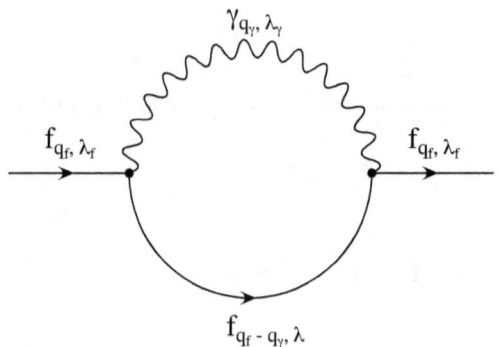

for a given mode $(\boldsymbol{q}_f, \lambda_f)$ of the fermion field.

8.9 Dressed photon

We consider a photon in the bare state $\left|1^{\gamma}_{\boldsymbol{q}_\gamma, \lambda_\gamma}\right\rangle$, where $\boldsymbol{q}_\gamma \neq \boldsymbol{0}$. Up to the first order, the corresponding dressed state is composed of the bare state, of states of the form $\left|1^{\gamma}_{\boldsymbol{q}_\gamma, \lambda_\gamma} 1^{\phi}_{\boldsymbol{q}-\boldsymbol{q}'_\gamma, \lambda'} 1^{\overline{\phi}}_{-\boldsymbol{q}, \lambda} 1^{\gamma}_{\boldsymbol{q}'_\gamma, \lambda'_\gamma}\right\rangle$, where ϕ is any electrically charged fermion, $\boldsymbol{q}'_\gamma \neq \boldsymbol{0}$ and $(\boldsymbol{q}'_\gamma, \lambda'_\gamma) \neq (\boldsymbol{q}_\gamma, \lambda_\gamma)$, of states of the form $\left|2^{\gamma}_{\boldsymbol{q}_\gamma, \lambda_\gamma} 1^{\phi}_{\boldsymbol{q}-\boldsymbol{q}_\gamma, \lambda'} 1^{\overline{\phi}}_{-\boldsymbol{q}, \lambda}\right\rangle$ as well as of states of the form $\left|1^{\phi}_{\boldsymbol{q}+\boldsymbol{q}_\gamma, \lambda'} 1^{\overline{\phi}}_{-\boldsymbol{q}, \lambda}\right\rangle$. The corresponding unnormalized coefficients are given by:

$$\tilde{\Phi}^{(1)} = -\sqrt{\frac{e^2}{4\pi\varepsilon_0 hc}}(1+2N)^{-3/2} Q_\phi \frac{u^{\phi\dagger}_{\boldsymbol{q}-\boldsymbol{q}'_\gamma, \lambda'} \gamma^0 \gamma u^{\overline{\phi}}_{-\boldsymbol{q}, \lambda} \cdot \varepsilon^*_{\boldsymbol{q}'_\gamma, \lambda'_\gamma}}{(2\pi q'_\gamma)^{1/2} \left(E^{\phi}_{\boldsymbol{q}-\boldsymbol{q}'_\gamma} + E^{\overline{\phi}}_{-\boldsymbol{q}} + E^{\gamma}_{\boldsymbol{q}'_\gamma}\right) a/hc}$$

for states of the form $\left| 1^\gamma_{\boldsymbol{q}_\gamma,\lambda_\gamma} 1^\phi_{\underline{\boldsymbol{q}-\boldsymbol{q}'_\gamma},\lambda'} 1^{\overline{\phi}}_{-\boldsymbol{q},\lambda} 1^\gamma_{\boldsymbol{q}'_\gamma,\lambda'_\gamma} \right\rangle$, by:

$$\widetilde{\Phi}^{(1)} = -\sqrt{2\frac{e^2}{4\pi\varepsilon_0 \hbar c}}(1+2N)^{-3/2}Q_\phi \frac{u^{\phi\dagger}_{\boldsymbol{q}-\boldsymbol{q}_\gamma,\lambda'}\gamma^0\gamma u^{\overline{\phi}}_{-\boldsymbol{q},\lambda}\cdot\boldsymbol{\varepsilon}^*_{\boldsymbol{q}_\gamma,\lambda_\gamma}}{(2\pi q_\gamma)^{1/2}\left(E^\phi_{\boldsymbol{q}-\boldsymbol{q}_\gamma}+E^{\overline{\phi}}_{-\boldsymbol{q}}+E^\gamma_{\boldsymbol{q}_\gamma}\right)}\,\mathrm{a/\hbar c}$$

for states of the form $\left| 2^\gamma_{\boldsymbol{q}_\gamma,\lambda_\gamma} 1^\phi_{\underline{\boldsymbol{q}-\boldsymbol{q}'_\gamma},\lambda'} 1^{\overline{\phi}}_{-\boldsymbol{q},\lambda} \right\rangle$, and by:

$$\widetilde{\Phi}^{(1)} = -\sqrt{\frac{e^2}{4\pi\varepsilon_0 \hbar c}}(1+2N)^{-3/2}Q_\phi \frac{u^{\phi\dagger}_{\boldsymbol{q}+\boldsymbol{q}_\gamma,\lambda'}\gamma^0\gamma u^{\overline{\phi}}_{-\boldsymbol{q},\lambda}\cdot\boldsymbol{\varepsilon}_{\boldsymbol{q}_\gamma,\lambda_\gamma}}{(2\pi q_\gamma)^{1/2}\left(E^\phi_{\boldsymbol{q}+\boldsymbol{q}_\gamma}+E^{\overline{\phi}}_{-\boldsymbol{q}}-E^\gamma_{\boldsymbol{q}_\gamma}\right)}\,\mathrm{a/\hbar c}$$

for states of the form $\left| 1^\phi_{\underline{\boldsymbol{q}+\boldsymbol{q}_\gamma},\lambda'} 1^{\overline{\phi}}_{-\boldsymbol{q},\lambda} \right\rangle$, respectively.

The corresponding energy is of second order and can be written as:

$$E^{(2)} = E^{(2)}(\Omega) - \frac{e^2}{4\pi\varepsilon_0 a}\sum_\phi Q_\phi^2 \kappa^\gamma_{\boldsymbol{q}_\gamma,\lambda_\gamma,\phi}$$

$$\kappa^\gamma_{\boldsymbol{q}_\gamma,\lambda_\gamma,\phi} := (1+2N)^{-3}\sum_q \frac{\sum_{\lambda',\lambda}\left|u^{\phi\dagger}_{\boldsymbol{q}-\boldsymbol{q}_\gamma,\lambda'}\gamma^0\gamma u^{\overline{\phi}}_{-\boldsymbol{q},\lambda}\cdot\boldsymbol{\varepsilon}^*_{\boldsymbol{q}_\gamma,\lambda_\gamma}\right|^2}{2\pi q_\gamma\left(E^\phi_{\boldsymbol{q}-\boldsymbol{q}_\gamma}+E^{\overline{\phi}}_{-\boldsymbol{q}}+E^\gamma_{\boldsymbol{q}_\gamma}\right)\mathrm{a/\hbar c}}$$

$$+(1+2N)^{-3}\sum_q \frac{\sum_{\lambda',\lambda}\left|u^{\phi\dagger}_{\boldsymbol{q}+\boldsymbol{q}_\gamma,\lambda'}\gamma^0\gamma u^{\overline{\phi}}_{-\boldsymbol{q},\lambda}\cdot\boldsymbol{\varepsilon}_{\boldsymbol{q}_\gamma,\lambda_\gamma}\right|^2}{2\pi q_\gamma\left(E^\phi_{\boldsymbol{q}+\boldsymbol{q}_\gamma}+E^{\overline{\phi}}_{-\boldsymbol{q}}-E^\gamma_{\boldsymbol{q}_\gamma}\right)\mathrm{a/\hbar c}}$$

where the spin summations evaluate to:

$$\sum_{\lambda',\lambda}\left|u^{\phi\dagger}_{\boldsymbol{q}-\boldsymbol{q}_\gamma,\lambda'}\gamma^0\gamma u^{\overline{\phi}}_{-\boldsymbol{q},\lambda}\cdot\boldsymbol{\varepsilon}^*_{\boldsymbol{q}_\gamma,\lambda_\gamma}\right|^2 = 1+\frac{M_\phi^2+(\boldsymbol{q}-\boldsymbol{q}_\gamma\cdot\boldsymbol{q}_\gamma)(\boldsymbol{q}\cdot\boldsymbol{q}_\gamma)/q_\gamma^2}{E^\phi_{\boldsymbol{q}-\boldsymbol{q}_\gamma}E^{\overline{\phi}}_{-\boldsymbol{q}}(\mathrm{a/\hbar c})^2}$$

$$\sum_{\lambda',\lambda}\left|u^{\phi\dagger}_{\boldsymbol{q}+\boldsymbol{q}_\gamma,\lambda'}\gamma^0\gamma u^{\overline{\phi}}_{-\boldsymbol{q},\lambda}\cdot\boldsymbol{\varepsilon}_{\boldsymbol{q}_\gamma,\lambda_\gamma}\right|^2 = 1+\frac{M_\phi^2+(\boldsymbol{q}+\boldsymbol{q}_\gamma\cdot\boldsymbol{q}_\gamma)(\boldsymbol{q}\cdot\boldsymbol{q}_\gamma)/q_\gamma^2}{E^\phi_{\boldsymbol{q}+\boldsymbol{q}_\gamma}E^{\overline{\phi}}_{-\boldsymbol{q}}(\mathrm{a/\hbar c})^2}$$

The vacuum energy diagram is also completed by adding following (negative) contribution:

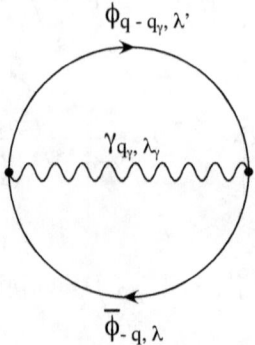

and by adding following (negative) self-energy diagram:

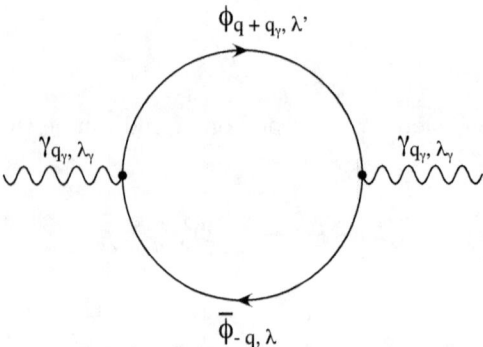

for a given mode $(\boldsymbol{q}_\gamma, \lambda_\gamma)$ of the photon field.

Part V

Examples

Chapter 9

Wave packets

9.1 Gaussian wave packet

A well-known consequence of the quantum formalism is the impossibility to describe a particle, like in classical mechanics, as a mass point having at each instant a well-defined position and velocity. In the quantum mechanics of a single particle in continuous space-time, the movement of the wave packet defining its statistical position can still be described, like in classical fluid mechanics, by a probability current density (which is related to the phase gradient of the wave packet), but as soon as several particles are present or are even being created and annihilated like in Quantum Field Theory, the analogy to classical fluid mechanics becomes much more elusive. It is still possible, however, to describe approximate particle trajectories in the frame of Quantum Field Theory if one considers that proper quantum effects may remain beyond the reach of experimental precision in some situations. Gaussian wave packets are a typical model of such particles with a quasi-classical behavior, *i.e.* with a position and a velocity being well-defined to a good approximation.

A Gaussian wave packet of a particle of type ϕ and in the spin state λ, with a mean momentum $\boldsymbol{q}_0 \in (\mathbb{Z}/(1+2\mathrm{N}))^3$, a mean position $\boldsymbol{n}_0 \in \mathbb{R}^3$ and a width $w_0 \in \mathbb{R}_+^*$, is given by:

$$|\Psi\rangle \;=\; \widehat{G_\lambda^\phi}^\dagger(\boldsymbol{q}_0, \boldsymbol{n}_0, w_0)\,|\Omega\rangle$$

$$\widehat{G_\lambda^\phi}^\dagger(\boldsymbol{q}_0, \boldsymbol{n}_0, w_0) \;:=\; C(\boldsymbol{q}_0, w_0)(2w_0)^{3/2}(1+2\mathrm{N})^{-3/2}$$

$$\sum_{\boldsymbol{q}} \exp\left(-2\pi w_0^2 \boldsymbol{q}^2 - \mathrm{i}2\pi\boldsymbol{q}\cdot\boldsymbol{n}_0\right) \widehat{a^\phi}^\dagger_{\boldsymbol{q}_0+\boldsymbol{q},\lambda}$$

with the normalization factor:

$$C(\boldsymbol{q}_0, w_0) := \left[(2w_0)^3(1+2\mathrm{N})^{-3}\sum_{\boldsymbol{q}}\exp\left(-4\pi w_0^2 \boldsymbol{q}^2\right)\right]^{-1/2}$$

In the usual case where $1 \ll w_0 \ll \mathrm{N}$, this normalization factor approximates to 1. On the position basis, the creation operator of the Gaussian wave packet can be expressed as:

$$\widehat{G^\phi_\lambda}^\dagger (\boldsymbol{q}_0, \boldsymbol{n}_0, w_0) \;=\; C(\boldsymbol{q}_0, w_0) w_0^{-3/2} \sum_{\boldsymbol{n}} A(\boldsymbol{n} - \boldsymbol{n}'_0(\boldsymbol{n}), w_0)$$

$$\exp\left(-\pi(\boldsymbol{n} - \boldsymbol{n}'_0(\boldsymbol{n}))^2/2w_0^2 + \mathrm{i}2\pi \boldsymbol{q}_0 \cdot \boldsymbol{n}\right) \widehat{a^\phi}^\dagger_{\boldsymbol{n},\lambda}$$

with the numerical factor:

$$A(\boldsymbol{n} - \boldsymbol{n}'_0(\boldsymbol{n}), w_0) := (2w_0^2)^{3/2}(1+2N)^{-3} \sum_{\boldsymbol{q}} \exp\left(-2\pi w_0^2 \left(\boldsymbol{q} - \mathrm{i}(\boldsymbol{n} - \boldsymbol{n}'_0(\boldsymbol{n}))/2w_0^2\right)^2\right)$$

where $\boldsymbol{n}'_0(\boldsymbol{n})$ can be chosen arbitrarily in $\boldsymbol{n}_0 + ((1 + 2N)\mathbb{Z})^3$. In the usual case where $1 \ll w_0 \ll N$, this factor approximates to 1 if $\boldsymbol{n}'_0(\boldsymbol{n})$ can be chosen such that $\|\boldsymbol{n} - \boldsymbol{n}'_0(\boldsymbol{n})\| \ll N$. To the zeroth order, the Hamiltonian evolution of the Gaussian wave packet $|\Psi_0\rangle = \widehat{G^\phi_\lambda}^\dagger (\boldsymbol{q}_0, \boldsymbol{n}_0, w_0) \, |\Omega\rangle$ is given by:

$$|\Psi_t\rangle \;=\; C(\boldsymbol{q}_0, w_0)(2w_0)^{3/2}(1 + 2N)^{-3/2}$$

$$\sum_{\boldsymbol{q}} \exp\left(-2\pi w_0^2 \boldsymbol{q}^2 - \mathrm{i}2\pi \boldsymbol{q} \cdot \boldsymbol{n}_0 - \mathrm{i}2\pi E^\phi_{\boldsymbol{q}_0+\boldsymbol{q}}(t - t_0)/\mathrm{h}\right) \widehat{a^\phi}^\dagger_{\boldsymbol{q}_0+\boldsymbol{q},\lambda} \, |\Omega\rangle$$

If $w_0 \gg q_0^{-1}$, the saddle-point approximation $E^\phi_{\boldsymbol{q}_0+\boldsymbol{q}} \approx E^\phi_{\boldsymbol{q}_0} + \boldsymbol{q} \cdot \nabla_{\boldsymbol{q}} E^\phi_{\boldsymbol{q}_0}$ can be used and it follows:

$$|\Psi_t\rangle \;\approx\; \exp\left(-\mathrm{i}2\pi E^\phi_{\boldsymbol{q}_0}(t - t_0)/\mathrm{h}\right) \widehat{G^\phi_\lambda}^\dagger (\boldsymbol{q}_0, \boldsymbol{n}_t, w_0) \, |\Omega\rangle$$

$$\boldsymbol{n}_t \;:=\; \boldsymbol{n}_0 + \boldsymbol{v}^\phi_{\boldsymbol{q}_0}(t - t_0)/a$$

The mean position $\boldsymbol{n}_t$ of the particle follows therefore, in the toroidal space $(\mathbb{R}/(1 + 2N)\mathbb{Z})^3$, a classical trajectory at the constant velocity $\boldsymbol{v}^\phi_{\boldsymbol{q}_0}$ which would be attributed classically to a point mass of mass m_ϕ and of momentum $\mathrm{h}\underline{\boldsymbol{q}}_0/a$.

Chapter 10

Coulomb scattering

10.1 Leading order calculation

We consider in this section the scattering of an electron by an atomic nucleus of atomic number Z. We model the nucleus by a classical point charge without magnetic moment, being at rest at the origin in the lattice reference frame and having a mass much higher than the mass of the electron. The corresponding electromagnetic field is described as a classical Coulomb potential V^{cl} given in terms of Fourier components by:

$$V_{\boldsymbol{n}}^{cl} := (1+2\mathrm{N})^{-3} \sum_{\boldsymbol{q}_\gamma \neq 0} \tilde{V}_{\boldsymbol{q}_\gamma}^{cl} \exp\left(\mathrm{i}2\pi\boldsymbol{n}\cdot\boldsymbol{q}_\gamma\right)$$

$$\tilde{V}_{\boldsymbol{q}_\gamma}^{cl} := \frac{Z\mathrm{e}}{4\pi^2\varepsilon_0 \mathrm{a} q_\gamma^2}$$

The corresponding semi-classical interaction Hamiltonian takes the form:

$$\widehat{\mathrm{H}}' := \widehat{\mathrm{H}}'_{QED} + \widehat{\mathrm{H}}^{cl}$$

$$\widehat{\mathrm{H}}^{cl} := \sum_{\boldsymbol{n}} \widehat{Q}_{\boldsymbol{n}} V_{\boldsymbol{n}}^{cl}$$

and its development on the plane wave basis is given by:

$$\widehat{\mathrm{H}}^{cl} = \frac{Z\mathrm{e}^2}{4\pi^2\varepsilon_0 \mathrm{a}}(1+2\mathrm{N})^{-3} \sum_{\phi,\boldsymbol{q},\lambda',\lambda} Q_\phi \sum_{\boldsymbol{q}_\gamma \neq 0} q_\gamma^{-2}$$

$$\left(\widehat{\overline{\psi^\phi}}_{\boldsymbol{q}+\boldsymbol{q}_\gamma,\lambda'} + \widehat{\overline{\psi^\phi}}_{-\boldsymbol{q}-\boldsymbol{q}_\gamma,\lambda'}\right) \gamma^0 \left(\widehat{\psi^\phi}_{\boldsymbol{q},\lambda} + \widehat{\overline{\psi}^\phi}_{-\boldsymbol{q},\lambda}\right)$$

As initial and final states, we take:

$$|\Psi_i\rangle := \left|1^e_{\boldsymbol{q}_i,\lambda_i}\right\rangle$$

$$|\Psi_f\rangle := \left|1^e_{\boldsymbol{q}_f,\lambda_f}\right\rangle$$

The matrix element of the interaction Hamiltonian is given for $q_f \neq q_i$ by:

$$H'_{f,i} = (1 + 2N)^{-3} \frac{-Ze^2}{4\pi^2\varepsilon_0 a\|q_f - q_i\|^2} u^{e\dagger}_{q_f,\lambda_f} u^e_{q_i,\lambda_i}$$

and the leading order transition probability for this process is also represented by following diagram:

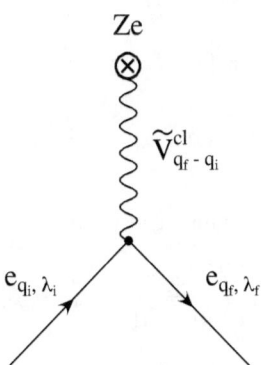

We consider a detector capturing the electrons having their momentum in the solid angle $\delta\Omega$. For $i \notin \delta\mathcal{F}$, the leading order transition probability takes the form:

$$\mathcal{P}^{(2)}\left(i \to \delta\mathcal{F}\right) \approx \int_{\delta\Omega} \int_0^\infty (2\pi)^2 \frac{t - t_0}{h} \left|H'_{f+\delta q,i}\right|^2 \delta^{(2)}_{t-t_0} \left(E_{f+\delta q} - E_i\right)$$

$$\left((1 + 2N)\frac{a}{h}\right)^3 p^2 dp d\Omega$$

By taking following continuation for the factors of the integrand:

$$\left|H'_{f+\delta q,i}\right|^2 \approx ((1 + 2N)a)^{-6} \frac{Z^2 e^4 h^4}{16\pi^4 \varepsilon_0^2 \|p - p_i\|^4} \left|u^{e\dagger}_{q,\lambda_f} u^e_{q_i,\lambda_i}\right|^2$$

$$\delta^{(2)}_{t-t_0} \left(E_{f+\delta q} - E_i\right) \approx \delta^{(2)}_{t-t_0} \left(\sqrt{(m_e c^2)^2 + (pc)^2} - E_i\right)$$

the integration over p yields to:

$$\mathcal{P}^{(2)}\left(i \to \delta\mathcal{F}\right) \approx (t - t_0) j_i \sigma^{(2)}\left(i \to \delta\mathcal{F}\right)$$

where j_i is the incident particle flux, given by:

$$j_i := ((1 + 2N)a)^{-3} v_i$$

$$v_i := \frac{p_i}{E_i/c^2}$$

and $\sigma^{(2)}\left(i \to \delta\mathcal{F}\right)$ the leading order cross section, given for $i \notin \delta\mathcal{F}$ by:

$$\sigma^{(2)}\left(i \to \delta\mathcal{F}\right) \approx \int_{\delta\Omega} \left(\frac{Ze^2}{8\pi\varepsilon_0 v_i p_i}\right)^2 \left|u^{e\dagger}_{q,\lambda_f} u^e_{q_i,\lambda_i}\right|^2 \frac{d\Omega}{\sin(\theta/2)^4}$$

where θ is the deviation angle of the electron.

If the incident electron beam isn't polarized and if the polarization of the scattered electron isn't being measured, the cross section is obtained by adding the cross sections corresponding to the final spin states λ_f and averaging over the cross sections corresponding to the initial spin states λ_i:

$$\left\langle \sigma^{(2)} \left(i \to \delta\mathcal{F}\right) \right\rangle \approx \int_{\delta\Omega} \left(\frac{Ze^2}{8\pi\varepsilon_0 v_i p_i}\right)^2 \frac{1}{2} \sum_{\lambda_f,\lambda_i} \left| u^{e\dagger}_{\boldsymbol{q},\lambda_f} u^{e}_{\boldsymbol{q}_i,\lambda_i} \right|^2 \frac{\mathrm{d}\Omega}{\sin\left(\theta/2\right)^4}$$

The spin summation is given for $E = E_i$ by:

$$\frac{1}{2} \sum_{\lambda_f,\lambda_i} \left| u^{e\dagger}_{\boldsymbol{q},\lambda_f} u^{e}_{\boldsymbol{q}_i,\lambda_i} \right|^2 = 1 - \beta_i^2 \sin\left(\theta/2\right)^2$$

and the mean cross section takes also the form:

$$\left\langle \sigma^{(2)} \left(i \to \delta\mathcal{F}\right) \right\rangle \approx \int_{\delta\Omega} \left(\frac{Ze^2}{8\pi\varepsilon_0 v_i p_i}\right)^2 \left(1 - \beta_i^2 \sin\left(\theta/2\right)^2\right) \frac{\mathrm{d}\Omega}{\sin\left(\theta/2\right)^4}$$

The total mean cross section for deviation angles $\theta \geq \theta_m$ is also given by:

$$\left\langle \sigma^{(2)} \left(i \to \mathcal{F}\right) \right\rangle \approx \left(\frac{Ze^2}{8\pi\varepsilon_0 v_i p_i}\right)^2 \left(\frac{4\pi}{\tan\left(\theta_m/2\right)^2} + 8\pi\beta_i^2 \ln\left(\sin\left(\theta_m/2\right)\right)\right)$$

Part VI

Appendix

Appendix A

Usual functions

A.1   The sinc function

In this document, the sinc function is defined by:

$$\operatorname{sinc}(X) := \begin{cases} 1 & \text{for } X = 0 \\ \sin(\pi X)/(\pi X) & \text{otherwise} \end{cases}$$

This function admits following integral expression:

$$\operatorname{sinc}(X) = \frac{1}{X} \int_{-X/2}^{X/2} \exp(i2\pi x)\, dx$$

and is normalized by:

$$\int_{-\infty}^{+\infty} \operatorname{sinc}(X)\, dX = 1$$

A.2   The esinc function

In this document, the esinc function is defined by:

$$\operatorname{esinc}(X) := \exp(i\pi X)\operatorname{sinc}(X)$$

where the sinc function is defined as in appendix A.1. This function admits following integral expression:

$$\operatorname{esinc}(X) = \frac{1}{X} \int_{0}^{X} \exp(i2\pi x)\, dx$$

can be written:

$$\operatorname{esinc}(X) = \frac{\sin(2\pi X)}{2\pi X} + i\frac{1 - \cos(2\pi X)}{2\pi X}$$

and verifies:

$$\operatorname{esinc}(-X) = \overline{\operatorname{esinc}(X)}$$

A.3 Nascent delta functions

In this document, we make use of following nascent delta functions, which converge to the delta energy distribution for $t - t_0 \to \infty$:

$$\delta_{t-t_0}^{(1)} (E) \quad := \quad \frac{t - t_0}{h} \text{sinc} \left(\frac{t - t_0}{h} E \right)$$

$$\delta_{t-t_0}^{(2)} (E) \quad := \quad \frac{t - t_0}{h} \text{sinc} \left(\frac{t - t_0}{h} E \right)^2$$

where the sinc function is defined as in appendix A.1. The square of the first one can be expressed in terms of the second one as:

$$\delta_{t-t_0}^{(1)} (E)^2 = \frac{t - t_0}{h} \delta_{t-t_0}^{(2)} (E)$$

We also make use of following family of functions converging to a distribution as $t - t_0 \to \infty$:

$$\delta_{t-t_0}^{(\text{P.V.})} (E) := 2 \frac{t - t_0}{h} \text{esinc} \left(\frac{t - t_0}{h} E \right)$$

where the esinc function is defined as in appendix A.2. Its limit is given by:

$$\lim_{t-t_0 \to \infty} \delta_{t-t_0}^{(\text{P.V.})} (E) = \delta(E) + \frac{i}{\pi} \text{P.V.} \left(\frac{1}{E} \right)$$

where the Cauchy principal value of $1/E$ is defined by its action on any test function $\phi(E)$ by:

$$\left(\text{P.V.} \left(\frac{1}{E} \right), \phi(E) \right) \quad := \quad \text{P.V.} \int_{-\infty}^{+\infty} \frac{\phi(E)}{E} dE$$

$$= \quad \lim_{\varepsilon \to 0^+} \left(\int_{-\infty}^{-\varepsilon} \frac{\phi(E)}{E} dE + \int_{\varepsilon}^{+\infty} \frac{\phi(E)}{E} dE \right)$$

Appendix B

Dirac and Pauli matrices

B.1 Pauli matrices

In this document, the Pauli matrices, which act canonically as endomorphisms of $\mathcal{H}^2$, are represented by:

$$\sigma_1 := \begin{pmatrix} 0 & 1 \\ 1 & 0 \end{pmatrix} \quad \sigma_2 := \begin{pmatrix} 0 & -i \\ i & 0 \end{pmatrix} \quad \sigma_3 := \begin{pmatrix} 1 & 0 \\ 0 & -1 \end{pmatrix}$$

These matrices verify the anticommutation relations:

$$\{\sigma_a, \sigma_b\} := \sigma_a \sigma_b + \sigma_b \sigma_a = 2\delta_{a,b} I_2$$

B.2 Dirac matrices

In this document, the Dirac matrices, which act canonically as endomorphisms of $\mathcal{H}^4$, are represented by:

$$\gamma^0 := \begin{pmatrix} I_2 & 0 \\ 0 & -I_2 \end{pmatrix} \quad \gamma^1 := \begin{pmatrix} 0 & \sigma_1 \\ -\sigma_1 & 0 \end{pmatrix}$$

$$\gamma^2 := \begin{pmatrix} 0 & \sigma_2 \\ -\sigma_2 & 0 \end{pmatrix} \quad \gamma^3 := \begin{pmatrix} 0 & \sigma_3 \\ -\sigma_3 & 0 \end{pmatrix}$$

These matrices verify the anticommutation relations:

$$\{\gamma^\mu, \gamma^\nu\} := \gamma^\mu \gamma^\nu + \gamma^\nu \gamma^\mu = 2g^{\mu\nu} I_4$$

We will make use of the condensed vectorial notation:

$$\boldsymbol{\gamma} := \begin{pmatrix} \gamma^1 \\ \gamma^2 \\ \gamma^3 \end{pmatrix}$$

Appendix C

Spinor operators

C.1   Photon spinor operators

In this document, we use following conventions for the polarization vectors of photons in the lattice reference frame:

$$\varepsilon_{q,1} \quad := \quad -\frac{1}{\sqrt{2}}\frac{1}{\sqrt{q_1^2 + q_2^2}}\frac{1}{q}\begin{pmatrix} q_1 q_3 - i q_2 q \\ q_2 q_3 + i q_1 q \\ -(q_1^2 + q_2^2) \end{pmatrix}$$

$$\varepsilon_{q,-1} \quad := \quad \frac{1}{\sqrt{2}}\frac{1}{\sqrt{q_1^2 + q_2^2}}\frac{1}{q}\begin{pmatrix} q_1 q_3 + i q_2 q \\ q_2 q_3 - i q_1 q \\ -(q_1^2 + q_2^2) \end{pmatrix}$$

For the special case of wave vectors q parallel to the third axis, we use the conventions:

$$\varepsilon_{q,1} \quad := \quad -\frac{1}{\sqrt{2}}\begin{pmatrix} 1 \\ i q_3/q \\ 0 \end{pmatrix}$$

$$\varepsilon_{q,-1} \quad := \quad \frac{1}{\sqrt{2}}\begin{pmatrix} 1 \\ -i q_3/q \\ 0 \end{pmatrix}$$

For the special case of the wave vector $q = 0$, we take $\varepsilon_{q,\lambda} := 0$. We extend this definition periodically to all $q \in \left(\frac{Z}{1+2N}\right)^3$ by $\varepsilon_{q,\lambda} := \varepsilon_{q,\lambda}$.

The polarization vectors of photons verify the Coulomb gauge conditions:

$$q \cdot \varepsilon_{q,\lambda} \quad = \quad 0$$
$$\varepsilon_{0,\lambda} \quad = \quad 0$$

as well as, for $q \neq 0$, the orthogonality relations:

$$\varepsilon_{q,\lambda'}^* \cdot \varepsilon_{q,\lambda} = \delta_{\lambda',\lambda}$$

One can also notice the relations:

$$\varepsilon_{-q,\lambda} = \varepsilon_{q,\lambda}^*$$
$$\varepsilon_{q,-\lambda} = -\varepsilon_{q,\lambda}^*$$

The photon annihilation and creation spinor operators, which act canonically as homomorphisms from $\mathcal{H}$ to $\mathcal{H}^3$, are defined for $q \neq 0$ by:

$$\widehat{\psi^\gamma}_{q,\lambda} := ((1+2N)a)^{-3/2} \sqrt{\frac{\mathrm{ha}}{8\pi^2\varepsilon_0 cq}} \, \varepsilon_{q,\lambda} \widehat{a^\gamma}_{q,\lambda} \sqrt{\widehat{N^\gamma}_{q,\lambda}}$$

$$\widehat{\psi^\gamma}^\dagger_{q,\lambda} := ((1+2N)a)^{-3/2} \sqrt{\frac{\mathrm{ha}}{8\pi^2\varepsilon_0 cq}} \, \varepsilon_{q,\lambda}^* \widehat{a^\gamma}^\dagger_{q,\lambda} \sqrt{1+\widehat{N^\gamma}_{q,\lambda}}$$

where ε_0 is the permittivity of the bare vacuum. For $q = 0$, we take $\widehat{\psi^\gamma}_{q,\lambda} = 0$ and $\widehat{\psi^\gamma}^\dagger_{q,\lambda} = 0$. The spinor operators can also be defined on the position basis by:

$$\widehat{\psi^\gamma}_{n,\lambda} := \sum_q \exp\left(\mathrm{i}2\pi n \cdot q\right) \widehat{\psi^\gamma}_{q,\lambda}$$

$$\widehat{\psi^\gamma}^\dagger_{n,\lambda} := \sum_q \exp\left(-\mathrm{i}2\pi n \cdot q\right) \widehat{\psi^\gamma}^\dagger_{q,\lambda}$$

We extend these definitions periodically to all $q \in \left(\frac{\mathbb{Z}}{1+2N}\right)^3$ by $\widehat{\psi^\gamma}_{q,\lambda} := \widehat{\psi^\gamma}_{\underline{q},\lambda}$ and $\widehat{\psi^\gamma}^\dagger_{q,\lambda} := \widehat{\psi^\gamma}^\dagger_{\underline{q},\lambda}$.

We will make use following condensed notation, representing a matrix acting canonically as an endomorphism of $\mathcal{H}^4$:

$$\gamma \cdot \varepsilon_{q,\lambda} := (\varepsilon_{q,\lambda})_1 \gamma^1 + (\varepsilon_{q,\lambda})_2 \gamma^2 + (\varepsilon_{q,\lambda})_3 \gamma^3$$

C.2 Fermion antisymmetrization operators

Let us define first a standard order on $\left(\frac{[-N,N]}{1+2N}\right)^3$, for instance the total order relation given by $q < q'$ if and only if one of the following assertions holds:

$$q_1 < q_1' \qquad \begin{cases} q_1 = q_1' \\ q_2 < q_2' \end{cases} \qquad \begin{cases} q_1 = q_1' \\ q_2 = q_2' \\ q_3 < q_3' \end{cases}$$

This allows us to label the particles of any field (ϕ, λ) present in a plane wave state $\left|(N_{q,\lambda}^\phi)\right\rangle$ in a standard way, using a standard particle numbering function π_λ^ϕ defined by:

$$\pi_\lambda^\phi\left((N_{q,\lambda}^\phi)\right) := (q_1, q_2, \ldots, q_{N_\lambda^\phi})$$

$$\begin{cases} q_1 \le q_2 \le \ldots \le q_{N_\lambda^\phi} \\ |\{i \mid q_i = q\}| = N_{q,\lambda}^\phi \end{cases}$$

The fermion antisymmetrization operators $\widehat{\epsilon^\phi}_{q,\lambda}$ can be defined conventionally with the help of this standard particle numbering function by their action on the momentum basis:

$$\widehat{\epsilon^\phi}_{q,\lambda} \left| (N_{q,\lambda}^\phi) \right\rangle := (-1)^\sigma \left| (N_{q,\lambda}^\phi) \right\rangle$$

$$\sigma = |\{i \mid q_i \le q\}|$$

$$(q_i) = \pi_\lambda^\phi \left((N_{q,\lambda}^\phi) \right)$$

These hermitian, unitary operators are always used together with the corresponding creation and annihilation operators for fermion fields. They verify following essential anticommutation properties, where the anticommutation notation $\{\cdot, \cdot\}$ is defined by $\{\widehat{a}, \widehat{b}\} := \widehat{a}\widehat{b} + \widehat{b}\widehat{a}$:

$$\{\widehat{\epsilon^\phi}_{q,\lambda}, \widehat{a^\phi}_{q,\lambda}\} = 0$$

$$\{\widehat{\epsilon^\phi}_{q,\lambda}\widehat{a^\phi}_{q,\lambda}, \widehat{\epsilon^\phi}_{q',\lambda}\widehat{a^\phi}_{q',\lambda}\} = 0$$

$$\{\widehat{a^\phi}^\dagger_{q,\lambda}\widehat{\epsilon^\phi}_{q,\lambda}, \widehat{a^\phi}^\dagger_{q',\lambda}\widehat{\epsilon^\phi}_{q',\lambda}\} = 0$$

$$\{\widehat{\epsilon^\phi}_{q,\lambda}\widehat{a^\phi}_{q,\lambda}, \widehat{a^\phi}^\dagger_{q',\lambda}\widehat{\epsilon^\phi}_{q',\lambda}\} = \delta_{q,q'}$$

C.3   Dirac spinor operators

In this document, we use following conventions for the Dirac spinors in the lattice reference frame (for charged leptons $\phi \in \{e, \mu, \tau\}$, neutrinos $\phi \in \{\nu_e, \nu_\mu, \nu_\tau\}$ and quarks $\phi \in \{u, c, t, d, s, b\}$):

$$u_{q,1/2}^\phi := \sqrt{\frac{1}{2}\left(1 + \frac{m_\phi c^2}{E}\right)} \begin{pmatrix} 1 \\ 0 \\ p_3/\left(m_\phi c + E/c\right) \\ (p_1 + ip_2)/\left(m_\phi c + E/c\right) \end{pmatrix}$$

$$u_{q,-1/2}^\phi := \sqrt{\frac{1}{2}\left(1 + \frac{m_\phi c^2}{E}\right)} \begin{pmatrix} 0 \\ 1 \\ (p_1 - ip_2)/\left(m_\phi c + E/c\right) \\ -p_3/\left(m_\phi c + E/c\right) \end{pmatrix}$$

$$u_{q,1/2}^{\overline{\phi}} := \sqrt{\frac{1}{2}\left(1 + \frac{m_\phi c^2}{E}\right)} \begin{pmatrix} (p_1 - ip_2)/\left(m_\phi c + E/c\right) \\ -p_3/\left(m_\phi c + E/c\right) \\ 0 \\ 1 \end{pmatrix}$$

$$u_{q,-1/2}^{\overline{\phi}} := \sqrt{\frac{1}{2}\left(1 + \frac{m_\phi c^2}{E}\right)} \begin{pmatrix} p_3/\left(m_\phi c + E/c\right) \\ (p_1 + ip_2)/\left(m_\phi c + E/c\right) \\ 1 \\ 0 \end{pmatrix}$$

In these expressions, we used the shorthand notations $E := E_q^\phi$ and $p := hq/a$ for energy and momentum. For the special case of the wave vector $q = 0$, we take the values:

$$u_{0,1/2}^\phi := \begin{pmatrix} 1 \\ 0 \\ 0 \\ 0 \end{pmatrix} \quad u_{0,-1/2}^\phi := \begin{pmatrix} 0 \\ 1 \\ 0 \\ 0 \end{pmatrix} \quad u_{0,1/2}^{\overline{\phi}} := \begin{pmatrix} 0 \\ 0 \\ 0 \\ 1 \end{pmatrix} \quad u_{0,-1/2}^{\overline{\phi}} := \begin{pmatrix} 0 \\ 0 \\ 1 \\ 0 \end{pmatrix}$$

as a definition for $m_\phi = 0$, too. We extend these definitions periodically to all $q \in \left(\frac{\mathbb{Z}}{1+2N}\right)^3$ by $u_{q,\lambda}^\phi := u_{\underline{q},\lambda}^\phi$ and $u_{q,\lambda}^{\overline{\phi}} := u_{\underline{q},\lambda}^{\overline{\phi}}$.

These spinors verify the orthogonality relations:

$$u_{q,\lambda'}^{\phi\,\dagger} u_{q,\lambda}^\phi = \delta_{\lambda',\lambda}$$

$$u_{q,\lambda'}^{\overline{\phi}\,\dagger} u_{q,\lambda}^{\overline{\phi}} = \delta_{\lambda',\lambda}$$

as well as the Dirac equations:

$$\gamma^\mu p_\mu u_{q,\lambda}^\phi = m_\phi c\, u_{q,\lambda}^\phi$$

$$\gamma^\mu p_\mu u_{q,\lambda}^{\overline{\phi}} = -m_\phi c\, u_{q,\lambda}^{\overline{\phi}}$$

where we use the condensed notation:

$$\gamma^\mu p_\mu := \frac{E}{c}\gamma^0 - p_1\gamma^1 - p_2\gamma^2 - p_3\gamma^3$$

The annihilation and creation spinor operators of these particles and of their anti-particles, which act canonically as homomorphisms between $\mathcal{H}$ and $\mathcal{H}^4$, are defined by:

$$\widehat{\psi^\phi}_{q,\lambda} := u_{q,\lambda}^\phi \widehat{\epsilon^\phi}_{q,\lambda} \widehat{a^\phi}_{q,\lambda}$$

$$\widehat{\overline{\psi}^\phi}_{q,\lambda} := u_{q,\lambda}^{\phi\,\dagger} \gamma^0 \widehat{a^\phi}_{q,\lambda}^\dagger \widehat{\epsilon^\phi}_{q,\lambda}$$

$$\widehat{\psi^{\overline{\phi}}}_{q,\lambda} := u_{q,\lambda}^{\overline{\phi}\,\dagger} \gamma^0 \widehat{\epsilon^{\overline{\phi}}}_{q,\lambda} \widehat{a^{\overline{\phi}}}_{q,\lambda}$$

$$\widehat{\overline{\psi}^{\overline{\phi}}}_{q,\lambda} := u_{q,\lambda}^{\overline{\phi}} \widehat{a^{\overline{\phi}}}_{q,\lambda}^\dagger \widehat{\epsilon^{\overline{\phi}}}_{q,\lambda}$$

where the fermion antisymmetrization operators $\widehat{\epsilon^\phi}_{q,\lambda}$ are defined as in appendix C.2. These spinor operators can also be defined on the position basis by:

$$\widehat{\psi^\phi}_{n,\lambda} := (1+2N)^{-3/2} \sum_q \exp\left(i2\pi n \cdot q\right) \widehat{\psi^\phi}_{q,\lambda}$$

$$\widehat{\overline{\psi}^\phi}_{n,\lambda} := (1+2N)^{-3/2} \sum_q \exp\left(-i2\pi n \cdot q\right) \widehat{\overline{\psi}^\phi}_{q,\lambda}$$

$$\widehat{\psi^{\overline{\phi}}}_{n,\lambda} := (1+2N)^{-3/2} \sum_q \exp\left(i2\pi n \cdot q\right) \widehat{\psi^{\overline{\phi}}}_{q,\lambda}$$

$$\widehat{\overline{\psi}^{\overline{\phi}}}_{n,\lambda} := (1+2N)^{-3/2} \sum_q \exp\left(-i2\pi n \cdot q\right) \widehat{\overline{\psi}^{\overline{\phi}}}_{q,\lambda}$$

We extend these definitions periodically to all $q \in \left(\frac{Z}{1+2N}\right)^3$ by $\widehat{\psi^\phi}_{q,\lambda} := \widehat{\psi^\phi}_{q,\lambda}$, $\widehat{\overline{\psi}^\phi}_{q,\lambda} := \widehat{\overline{\psi}^\phi}_{q,\lambda}$, $\widehat{\psi^{\overline{\phi}}}_{q,\lambda} := \widehat{\psi^{\overline{\phi}}}_{q,\lambda}$ and $\widehat{\overline{\psi}^{\overline{\phi}}}_{q,\lambda} := \widehat{\overline{\psi}^{\overline{\phi}}}_{q,\lambda}$.

Spinor products

In the development of the interaction Hamiltonian on the plane waves basis, the Dirac spinors always appear in the form of products. For instance (see section 8.5), the elastic scattering terms of the Coulomb interaction, with or without spin flip, contain following products, where we use the shorthand notations $\gamma := \gamma_q^\phi$ and $\gamma' := \gamma_{q'}^\phi$ for the Lorentz factors:

$$u^{\phi\dagger}_{q',\pm 1/2} u^{\phi}_{q,\pm 1/2} = \frac{1}{2}\sqrt{1+\frac{1}{\gamma'}}\sqrt{1+\frac{1}{\gamma}}\left(1 + \frac{q' \cdot q \pm i(q' \times q)_3}{M_\phi^2(1+\gamma')(1+\gamma)}\right)$$

$$u^{\phi\dagger}_{q',\mp 1/2} u^{\phi}_{q,\pm 1/2} = \frac{i(q' \times q)_1 \mp (q' \times q)_2}{2M_\phi^2\sqrt{(1+\gamma')(1+\gamma)\gamma'\gamma}}$$

These expressions are also valid for the corresponding antiparticles $\overline{\phi}$. The particle pair creation and annihilation terms of the Coulomb interaction, with equal or opposite spins, contain following products:

$$u^{\phi\dagger}_{q',\pm 1/2} u^{\overline{\phi}}_{q,\pm 1/2} = \frac{1}{2}\sqrt{1+\frac{1}{\gamma'}}\sqrt{1+\frac{1}{\gamma}}\left(\frac{q'_1 \mp iq'_2}{M_\phi(1+\gamma')} + \frac{q_1 \mp iq_2}{M_\phi(1+\gamma)}\right)$$

$$u^{\phi\dagger}_{q',\mp 1/2} u^{\overline{\phi}}_{q,\pm 1/2} = \frac{1}{2}\sqrt{1+\frac{1}{\gamma'}}\sqrt{1+\frac{1}{\gamma}}\left(\frac{\mp q'_3}{M_\phi(1+\gamma')} + \frac{\mp q_3}{M_\phi(1+\gamma)}\right)$$

The calculation of transition probabilities involves the square of the absolute value of these products, given by:

$$\left|u^{\phi\dagger}_{q',\lambda} u^{\phi}_{q,\lambda}\right|^2 = \frac{(q' \cdot q)^2 + (q' \times q)_3^2}{4M_\phi^4(1+\gamma')(1+\gamma)\gamma'\gamma} + \frac{1}{4\gamma'\gamma}\left((1+\gamma')(1+\gamma) + \frac{2q' \cdot q}{M_\phi^2}\right)$$

$$\left|u^{\phi\dagger}_{q',-\lambda} u^{\phi}_{q,\lambda}\right|^2 = \frac{(q' \times q)_1^2 + (q' \times q)_2^2}{4M_\phi^4(1+\gamma')(1+\gamma)\gamma'\gamma}$$

Their sum, involved when the spin of the scattered particles isn't being measured, takes the form:

$$\sum_{\lambda'}\left|u^{\phi\dagger}_{q',\lambda'} u^{\phi}_{q,\lambda}\right|^2 = \frac{1}{4}\left(\sqrt{1+\frac{1}{\gamma'}}\sqrt{1+\frac{1}{\gamma}} + \sqrt{1-\frac{1}{\gamma'}}\sqrt{1-\frac{1}{\gamma}}\right)^2 - \beta'\beta\sin\left(\theta/2\right)^2$$

where we used the shorthand notations $\beta := \beta_q^\phi$ and $\beta' := \beta_{q'}^\phi$ and where we introduced the scattering angle θ between q and q'. In the case where this angle is undefined because one of q or q' is zero, the last term can be dropped.

Bibliography

[1] Adams, D. (1979). *The Hitchhiker's Guide to the Galaxy*. Pan Books, London (United Kingdom), 1979.

[2] Aspect, A.; et al. (1982). *Experimental Realization of Einstein-Podolsky-Rosen-Bohm Gedankenexperiment: A New Violation of Bell's Inequalities*. Physical Review Letters 49: 91, American Physical Society, College Park (Maryland, United States), 1982.

[3] Bassi, A.; Ghirardi, G. (2003). *Dynamical Reduction Models*. Physics Reports 379: 257, Elsevier, Amsterdam (The Netherlands), 2003.

[4] CMS Collaboration (2011). *Measurement of W^+W^- Production and Search for the Higgs Boson in pp Collisions at $\sqrt{s} = 7$ TeV*. Physics Letters B 699: 25, Elsevier, Amsterdam (The Netherlands), 2011.

[5] Collins, J. (1984). *Renormalization: an introduction to renormalization, the renormalization group and the operator product expansion*. Cambridge University Press, Cambridge (United Kingdom), 1984.

[6] Descartes, R. (1637). *Discours de la méthode pour bien conduire sa raison, et chercher la vérité dans les sciences*. Ian Maire, Leiden (The Netherlands), 1637. *Discourse on the Method of Rightly Conducting One's Reason and of Seeking Truth in the Sciences*. W. Blackwood and sons, London (United Kingdom), 1870.

[7] Einstein, A. (1916). *Über die spezielle und die allgemeine Relativitätstheorie*. Vieweg, Braunschweig (Germany), 1917. *Relativity: The Special and General Theory*. Methuen & Co. Ltd., London (United Kingdom), 1920.

[8] Fauvel, S. (2010). *Quantum Ethics: A Spinozist Interpretation of Quantum Field Theory*. CreateSpace Independent Publishing Platform, Scotts Valley (California, United States), 2013.

[9] Feynman, R. (1964). *The Character of Physical Law: 6. Probability and uncertainty - the quantum mechanical view of nature*. Messenger Lectures, Cornell University, Ithaca (New York, United States), 1964. The British Broadcasting Corporation, London (United Kingdom), 1965.

[10] Freedman, W. L.; et al. (2001). *Final Results from the Hubble Space Telescope Key Project to Measure the Hubble Constant*. The Astrophysical Journal 553 (1): 47–72, IOP Publishing, Bristol (United Kingdom), 2001.

[11] Friedman, A. (1922). *Über die Krümmung des Raumes*. Zeitschrift für Physik A 10 (1): 377–386, Springer, Berlin (Germany), 1922.

[12] Hermanns, W. (1983). *Einstein and the Poet: In Search of the Cosmic Man*. Branden Press, Boston (Massachusetts, United States), 1983.

[13] Kogut, A.; et al. (1993). *Dipole Anisotropy in the COBE Differential Microwave Radiometers First-Year Sky Maps*. The Astrophysical Journal 419: 1–6, IOP Publishing, Bristol (United Kingdom), 1993.

[14] Komatsu, E.; et al. (2010). *Seven-Year Wilkinson Microwave Anisotropy Probe (WMAP) Observations: Cosmological Interpretation*. The Astrophysical Journal Supplement Series 192: 18, IOP Publishing, Bristol (United Kingdom), 2011.

[15] Marin, J. M. (2009). *'Mysticism' in quantum mechanics: the forgotten controversy*. European Journal of Physics 30: 807-822, IOP Publishing, Bristol (United Kingdom), 2009.

[16] Neumann, J. v. (1932). *Mathematische Grundlagen der Quantenmechanik*. Springer, Berlin (Germany), 1932. *Mathematical Foundations of Quantum Mechanics*. Princeton University Press, Princeton (New Jersey, United States), 1955.

[17] Rabin, J. M. (1981). *Long Range Interactions In Lattice Field Theory*. Ph.D. Thesis SLAC-0240, Stanford (California, United States), 1981.

[18] Spinoza, B. (1677). *Ethica, Ordine Geometrico demonstrata*. Amsterdam (The Netherlands), 1677. *The Ethics*. London (United Kingdom), 1883.

www.ingramcontent.com/pod-product-compliance
Lightning Source LLC
Chambersburg PA
CBHW071722170526
45165CB00005B/2116